Understanding Alzheimer's Disease

Understanding Alzheimer's Disease

A Comprehensive Q&A Guide

Austin Mardon, PhD

Mercury Learning and Information
BOSTON, MASSACHUSETTS

MERCURY LEARNING AND INFORMATION
121 High Street, 3rd Floor
Boston, MA 02110
info@merclearning.com

A. Mardon. *Understanding Alzheimer's Disease: A Comprehensive Q&A Guide.*
ISBN: 978-1-5015-2746-3

Library of Congress Control Number: 2025949533

242526321 This book is printed on acid-free paper in the United States of America.

Contents

Preface

Alzheimer's disease is a complex and challenging condition that affects millions of people worldwide. As a leading cause of dementia, it poses significant emotional, physical, and financial burdens on patients, caregivers, and society. This book aims to provide a comprehensive overview of Alzheimer's disease, addressing its various aspects from early detection and diagnosis to caregiving and emerging treatments. By offering clear, informative answers to common questions, we hope to empower readers with the knowledge they need to navigate the complexities of this condition and support those affected by it.

Introduction to Alzheimer's Disease

CHAPTER 1

Welcome to the chapter "Introduction to Alzheimer's Disease," where we embark on a journey to understand one of the most prevalent neurodegenerative disorders affecting millions worldwide. Alzheimer's disease is characterized by progressive cognitive decline, memory loss, and behavioral changes that significantly impact daily functioning and quality of life. As our understanding of Alzheimer's continues to evolve, so does our approach to diagnosis, treatment, and caregiving.

In this Q&A chapter, we will explore the fundamentals of Alzheimer's disease, including its symptoms, risk factors, and underlying causes. We will discuss the importance of early detection and diagnosis, the current state of research into potential treatments and preventive measures and provide practical advice for caregivers and families navigating the challenges of Alzheimer's. Our aim is to deliver clear, evidence-based information that empowers individuals affected by Alzheimer's disease and their support networks to make informed decisions and foster a compassionate and supportive environment.

Q1: What is Alzheimer's disease?

Alzheimer's disease is a progressive neurodegenerative disorder that primarily affects memory and cognitive function. It is characterized by the accumulation of amyloid plaques and tau tangles in the brain, leading to the death of brain cells and the deterioration of mental faculties. This condition typically manifests in individuals over the age of 65, but early-onset forms can occur in younger individuals. Alzheimer's disease progresses through various stages, starting with mild memory lapses and advancing to severe cognitive and functional impairments, ultimately resulting in the need for full-time care.

Q2: Who first discovered Alzheimer's disease, and in what year was it identified?

In 1901, Dr. Alois Alzheimer, a German psychiatrist and neuropathologist, began studying a 51-year-old patient named

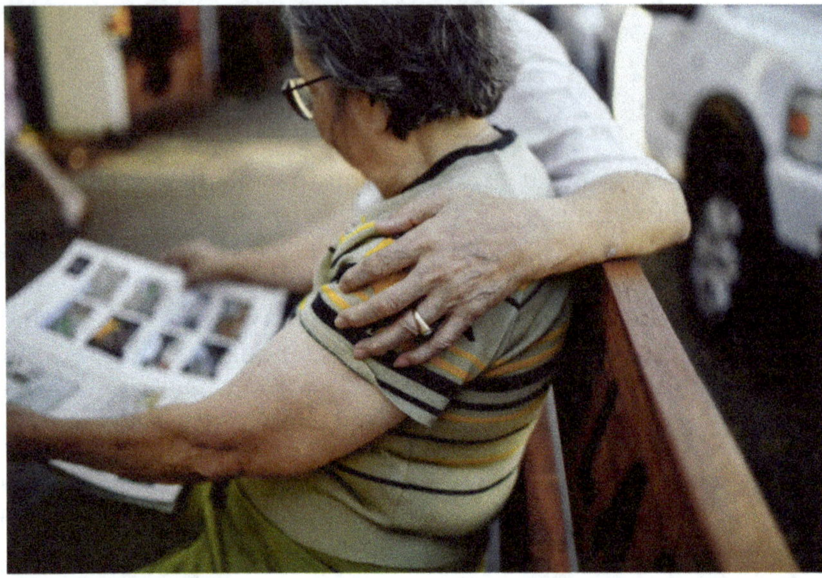

▲ **FIGURE 1.1.**
Couple reading the newspaper.

Auguste Deter at the Frankfurt Asylum. Auguste exhibited severe memory loss, language difficulties, disorientation, and behavioral changes, including paranoia and mood swings—symptoms highly unusual for someone her age. She frequently forgot recent conversations, lost track of time, and accused others of plotting against her. Strikingly aware of her decline, she once told Dr. Alzheimer, "I have lost myself," poignantly reflecting the distress and identity loss caused by her condition.

Dr. Alzheimer meticulously documented her symptoms over several years. When Auguste passed away in 1906, he performed an autopsy and discovered two distinctive brain abnormalities: amyloid plaques and neurofibrillary tangles. These findings were groundbreaking, linking physical brain changes to behavioral symptoms.

Later in 1906, Dr. Alzheimer presented his findings at a medical conference in Tübingen, Germany, and published his work shortly afterward. Initially referred to as *presenile dementia*, this discovery laid the foundation for identifying what is now known as Alzheimer's disease, marking a pivotal moment in the study of dementia.

▲ FIGURE 1.2.
Alzheimer's can have a big impact on family life, but with patience and understanding love always shines through.

Q3: What are amyloid plaques and tau tangles?

Amyloid plaques and tau tangles are hallmark pathological features of Alzheimer's disease. Amyloid plaques are abnormal clumps of protein fragments called beta-amyloid that accumulate between nerve cells, disrupting their function. Tau tangles, on the other hand, are twisted fibers of a protein called tau that form inside dying cells, interfering with the transport of nutrients and other essential substances. Together, these abnormalities contribute to the degeneration of brain cells and the progression of Alzheimer's disease.

Q4: What are the early signs and symptoms of Alzheimer's disease?

The early signs of Alzheimer's disease often include subtle memory loss, particularly short-term memory issues, such as forgetting recent conversations or misplacing items. Individuals may also have trouble with problem-solving, language, and completing familiar tasks. Other early symptoms include confusion about time and place, decreased judgment, and

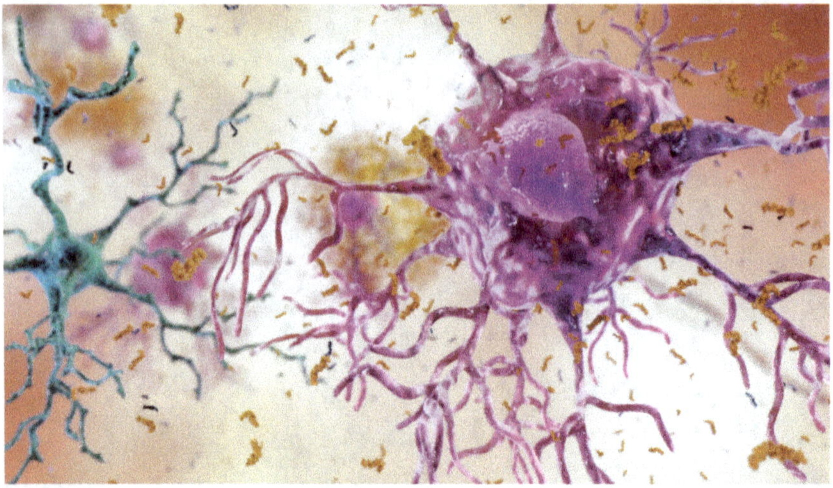

▲ **FIGURE 1.3**
Brain scan image showing Alzheimer's effects.

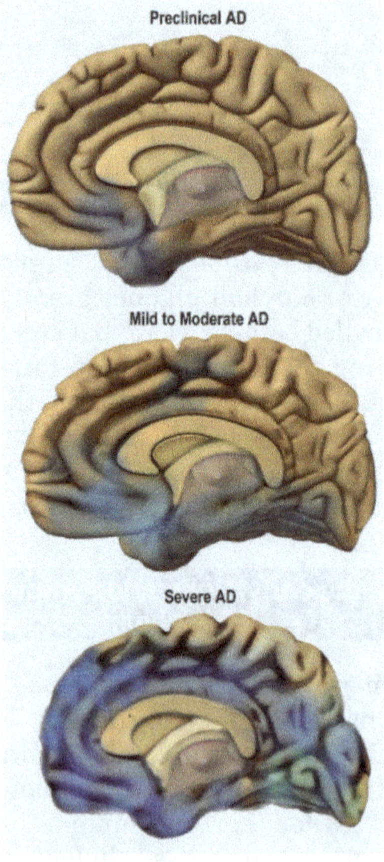

▲ **FIGURE 1.4**
Mild-severe AD brain changes.

changes in mood or personality. Recognizing these early signs is crucial for timely diagnosis and intervention, as early-stage Alzheimer's can sometimes be mistaken for normal aging.

Q5: How common is Alzheimer's disease?

Alzheimer's disease is the most common form of dementia, accounting for 60–80% of all dementia cases. It affects millions of people worldwide and is a significant public health concern. The risk of developing Alzheimer's increases with age, and with the global population aging, the prevalence of the disease is expected to rise substantially in the coming decades.

Q6: What is the global impact of Alzheimer's disease?

Alzheimer's disease is a global health challenge, affecting millions of people across different cultures and regions. The increasing prevalence of Alzheimer's due to aging populations poses significant challenges for healthcare systems worldwide. Efforts to address Alzheimer's include research initiatives, public health campaigns, and policy development aimed at improving diagnosis, care, and support for individuals and families affected by the disease.

Q7: Can Alzheimer's disease be prevented?

Although there is no guaranteed way to prevent Alzheimer's disease, research suggests that certain lifestyle factors may help reduce the risk. These include maintaining a healthy diet, staying physically active, engaging in regular cognitive and social activities, managing cardiovascular health, and avoiding smoking. Such measures support overall brain health and may lower the risk of developing Alzheimer's and other dementias.

Q8: How was Alzheimer's coined?

The term *Alzheimer's disease* was coined by Dr. Emil Kraepelin, a German psychiatrist and colleague of Dr. Alois Alzheimer. Kraepelin first used the term in 1910 in the eighth edition of his influential *Textbook of Psychiatry*, where he described the disease based on the symptoms and brain abnormalities documented by Dr. Alzheimer in his patient, Auguste Deter.

Kraepelin, a key figure in early psychiatric research, believed that Dr. Alzheimer's findings indicated a unique form of

▲ **FIGURE 1.5**
Alois Alzheimer.

dementia distinct from other mental illnesses or the common forms of senility associated with aging. Recognizing the significance of his colleague's work, Kraepelin named the condition Alzheimer's disease to honor Dr. Alzheimer's contributions.

This naming marked a pivotal shift in understanding dementia, as it began to distinguish specific causes of dementia rather than attributing all cases to generalized aging. Over time, Alzheimer's disease became recognized as a specific progressive neurodegenerative condition, characterized by memory loss, cognitive decline, and hallmark changes in the brain, such as amyloid plaques and tau tangles.

Q9: How does Alzheimer's disease affect the brain?

Alzheimer's disease progressively damages the brain by causing neuronal death and disrupting neural connections. Key areas affected include the hippocampus, essential for memory formation, and the cerebral cortex, responsible for higher-order cognitive functions like thinking, planning, and decision-making.

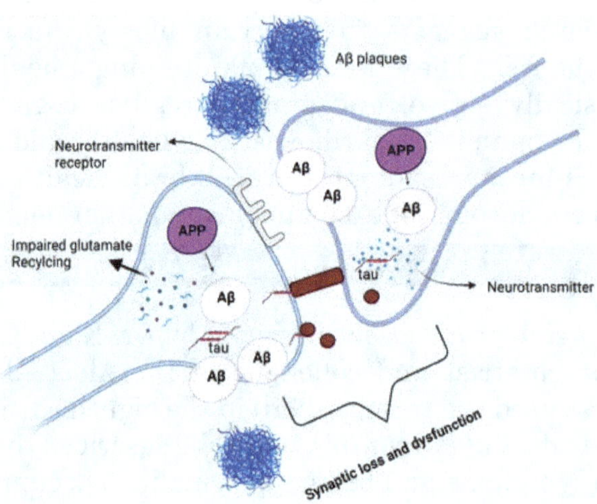

▲ **FIGURE 1.6**
How the tangled proteins in the brain react with neurons to cause the physiological state of Alzheimer's disease.

As the disease progresses, widespread brain atrophy occurs, resulting in significant cognitive and functional impairments.

Q10: How is Alzheimer's disease different from normal aging?

Normal aging involves some degree of cognitive decline, such as slower processing speeds and occasional memory lapses. However, these changes do not interfere significantly with daily functioning. In contrast, Alzheimer's disease involves more severe and progressive cognitive decline that disrupts daily life and activities. Distinguishing between normal aging and early signs of Alzheimer's is crucial for timely diagnosis and intervention.

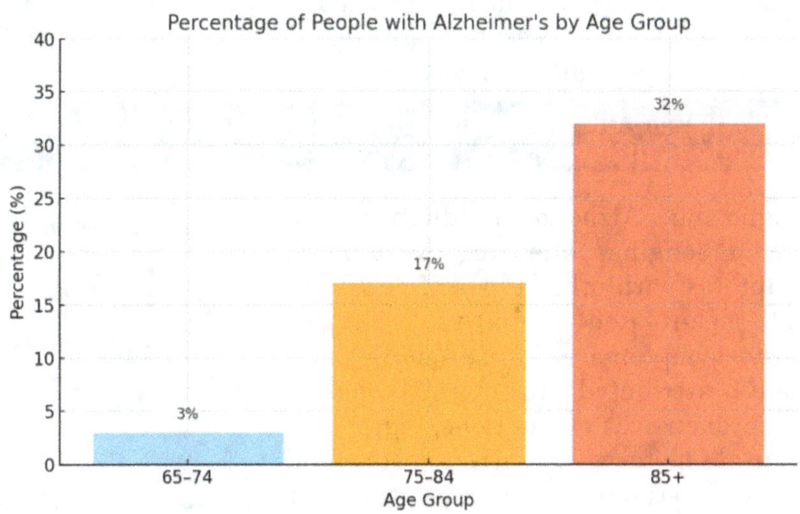

▲ **FIGURE 1.7**
Chart showing Alzheimer's prevalence by age group.

Q11: Are there different types of Alzheimer's disease?

Alzheimer's disease can be classified into two main types: early-onset and late-onset. *Early-onset Alzheimer's* occurs in individuals under the age of 65 and is often linked to genetic mutations. *Late-onset Alzheimer's*, the more common form, typically develops after age 65 and is influenced by a combination of genetic, environmental, and lifestyle factors. Both types share similar symptoms and pathological features but may differ in their underlying causes and progression.

Q12: How does Alzheimer's disease differ from other types of dementia?

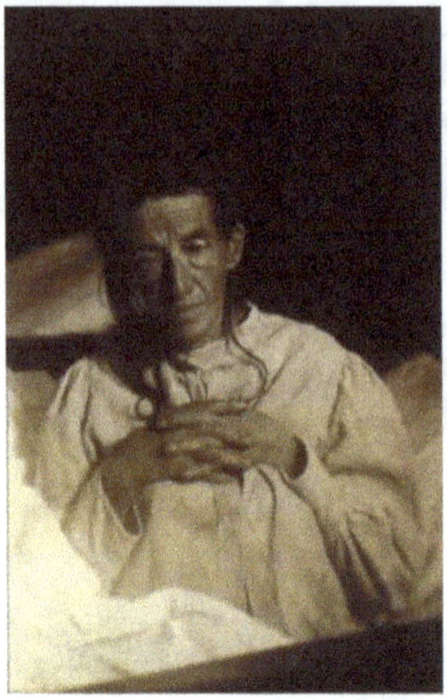

▲ **FIGURE 1.8**
Auguste Deter.

While Alzheimer's disease is the most common cause of dementia, it is not the only type. Dementia is an umbrella term for a range of conditions that cause cognitive decline, including vascular dementia, Lewy body dementia, and frontotemporal dementia. Each type has distinct characteristics: For example, vascular dementia often follows strokes and affects planning and judgment, while Lewy body dementia includes symptoms like visual hallucinations and movement issues. Alzheimer's disease is primarily marked by memory loss and progressive cognitive decline due to specific brain changes.

Q13: What are the challenges of diagnosing Alzheimer's disease?

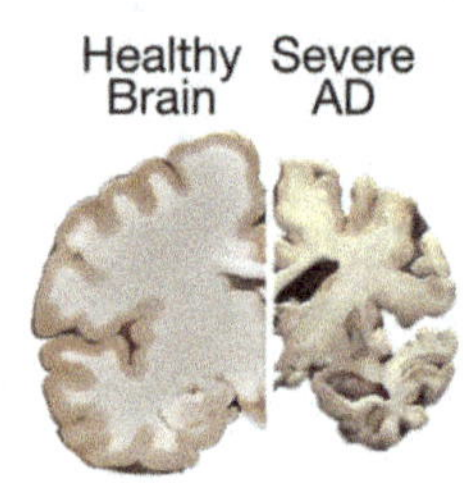

▲ **FIGURE 1.9**
Healthy brain vs. severe AD brain.

Diagnosing Alzheimer's disease can be challenging due to overlapping symptoms with other types of dementia and age-related cognitive changes. Early symptoms may be subtle and easily attributed to normal aging. Comprehensive evaluations, including medical history, cognitive tests, and imaging studies, are essential for an accurate diagnosis. Advances in biomarkers and diagnostic technologies are improving the ability to detect Alzheimer's disease earlier and with greater accuracy.

Q14: What role does inflammation play in Alzheimer's disease?

Inflammation is believed to play a significant role in the development and progression of Alzheimer's disease. Chronic inflammation in the brain, often triggered by the buildup of amyloid plaques, can lead to neuronal damage and worsen the disease's pathology. Ongoing research aims to understand the mechanisms of inflammation in Alzheimer's and develop anti-inflammatory treatments that could potentially slow its progression.

Q15: How is Alzheimer's disease linked to Down syndrome?

Individuals with Down syndrome have an increased risk of developing Alzheimer's disease, often at an earlier age than the general population. This increased risk is linked to the presence of an extra copy of chromosome 21, which carries the gene for amyloid precursor protein (APP). The overproduction of APP leads to the accumulation of amyloid plaques in the brain, contributing to the development of Alzheimer's disease.

Q16: How do comorbidities affect Alzheimer's disease progression?

Comorbidities, or coexisting medical conditions, can complicate the management and progression of Alzheimer's disease. Conditions such as diabetes, hypertension, heart disease, and depression may accelerate cognitive decline and increase the burden of care. Effectively managing comorbidities through medication, lifestyle changes, and regular medical checkups is essential for enhancing the quality of life for individuals with Alzheimer's.

Q17: What is the role of oxidative stress in Alzheimer's disease?

Oxidative stress, defined as an imbalance between the production of free radicals (unstable molecules that can damage cells) and the body's ability to neutralize them with antioxidants, is believed to contribute to the development and progression of Alzheimer's disease. Free radicals can damage neurons and other brain cells, impairing their function and triggering cell death, which accelerates cognitive decline. Research is exploring the potential of antioxidants and other compounds to counteract oxidative stress, reduce cellular damage, and protect brain health in individuals with Alzheimer's.

Q18: How does Alzheimer's disease affect families?

Alzheimer's disease profoundly affects families, placing emotional, physical, and financial burdens on caregivers. Family members often take on caregiving roles, which can lead to stress, burnout, and strained relationships. The progressive nature of the disease requires increasing levels of care and support, impacting family dynamics and requiring significant adjustments to daily life.

Q19: How does Alzheimer's disease impact relationships?

Alzheimer's disease can strain relationships as roles and dynamics change. Spouses, children, and friends may struggle with the emotional toll of seeing their loved one decline. Caregivers often experience stress and burnout, which can affect their own health and well-being. Open communication, seeking support, and finding ways to maintain meaningful connections are essential for preserving relationships and providing effective care.

Q20: How does Alzheimer's disease affect daily life?

As Alzheimer's disease progresses, it increasingly impacts daily life and independence. In the early stages, individuals may struggle with tasks that require memory and organization, such as managing finances or following a recipe. In the moderate to severe stages, basic *activities of daily living* (ADLs), such as dressing, bathing, and eating, become challenging and require assistance from caregivers. Maintaining a structured and supportive environment can help manage these daily challenges.

Q21: What are the psychological and emotional effects of Alzheimer's disease?

Alzheimer's disease can have profound psychological and emotional effects on patients and their families. Patients may experience depression, anxiety, frustration, and confusion as their cognitive abilities decline. Caregivers often face significant emotional stress, burnout, and feelings of grief as they watch their loved one's condition worsen. Support, counseling, and coping strategies are crucial for managing the emotional impact of Alzheimer's disease.

Q22: What is the impact of Alzheimer's disease on personal identity?

Alzheimer's disease can profoundly impact an individual's sense of personal identity, as memory loss and cognitive decline affect self-awareness and the ability to recall life experiences. This loss of identity can be distressing for both patients and their families. Supporting individuals with Alzheimer's involves recognizing and respecting their remaining abilities and interests, fostering a sense of continuity and dignity.

Q23: How does Alzheimer's disease affect the senses?

Alzheimer's disease can impact the senses, including vision, hearing, taste, smell, and touch. Sensory changes may result in difficulties with perception, balance, and coordination, often leading to disorientation and increased anxiety. Engaging in activities that stimulate the senses and ensuring regular checkups for vision and hearing can help manage these effects.

Q24: How does Alzheimer's disease affect communication?

Alzheimer's disease progressively impairs communication skills, making it increasingly difficult for individuals to express themselves and understand others. In the early stages, individuals may struggle to find words and follow conversations, while in the later stages, they may have minimal or no verbal communication. Caregivers can use strategies such as speaking slowly, using simple language, and incorporating nonverbal cues to facilitate communication and reduce frustration.

Q25: How does Alzheimer's disease affect eating and nutrition?

Alzheimer's disease can impact eating and nutrition by causing changes in appetite, difficulty swallowing, and forgetting to eat. Individuals may also have altered taste preferences and difficulty recognizing food. Ensuring proper nutrition involves offering balanced meals, creating a calm dining environment, and helping as needed. Nutritional supplements and modified food textures can help address specific challenges.

Q26: How does Alzheimer's disease affect decision-making and judgment?

Alzheimer's disease impairs the brain's ability to process information and make decisions, leading to difficulties with judgment and problem-solving. Individuals may struggle to evaluate options, make informed choices, and understand the consequences of their actions. This decline in executive function can impact daily activities, financial management, and safety. Providing support and supervision is essential to help individuals manage these challenges.

Q27: What role does diet play in Alzheimer's disease?

Diet plays a crucial role in brain health and can influence the risk and progression of Alzheimer's disease. Nutrient-rich diets, such as the Mediterranean diet, which includes fruits, vegetables, whole grains, lean proteins, and healthy fats, have been linked to a lower risk of Alzheimer's. Specific nutrients like omega-3 fatty acids, antioxidants, and vitamins are thought to support cognitive function and protect against neurodegeneration.

Q28: What is the role of physical activity in Alzheimer's disease?

Regular physical activity benefits overall health and may help reduce the risk and progression of Alzheimer's disease. Exercise supports cardiovascular health, enhances mood, and promotes cognitive function. Activities like walking, swimming, and strength training can be tailored to different fitness levels and abilities, making physical activity an essential component of Alzheimer's disease management.

Q29: How does social engagement impact Alzheimer's disease?

Social engagement is crucial for cognitive health and can help delay the onset and progression of Alzheimer's disease. Staying socially active through interactions with family, friends, and community groups provides mental stimulation and emotional support. Social activities, such as group exercise, volunteering, attending events, or participating in clubs, contribute to a sense of purpose and well-being.

Q30: What is the role of cognitive stimulation in Alzheimer's disease?

Cognitive stimulation involves activities that challenge the brain and promote mental agility, such as puzzles, games, reading, and learning new skills. Engaging in cognitive stimulation can help maintain cognitive function and potentially slow the progression of Alzheimer's disease. Cognitive training and rehabilitation programs are valuable tools for supporting individuals with Alzheimer's.

Q31: What is the role of technology in managing Alzheimer's disease?

Technology plays an increasingly important role in managing Alzheimer's disease. Tools like medication reminders, GPS trackers, and smart home devices enhance safety and independence. Cognitive training apps and virtual reality experiences provide engaging ways to stimulate the brain. Additionally, telemedicine offers convenient access to healthcare professionals and support services, making it easier for individuals with Alzheimer's and their caregivers to receive care and guidance.

Q32: How does Alzheimer's disease impact mobility and physical function?

Alzheimer's disease can affect mobility and physical function, especially in the later stages. Individuals may experience difficulties with coordination, balance, and motor skills, increasing the risk of falls and injuries. Physical therapy and regular exercise can help maintain mobility and strength, while assistive devices, such as walkers and handrails, can enhance safety and independence.

Q33: How does Alzheimer's disease affect financial management?

Alzheimer's disease can significantly impair an individual's ability to manage finances, including paying bills, budgeting, and making financial decisions. Early in the disease, individuals may make errors or forget payments, while more advanced stages require complete oversight by caregivers or financial representatives. Establishing legal and financial plans early, such as power of attorney and trusts, can help protect assets and ensure financial stability.

Q34: What are the ethical considerations in Alzheimer's disease care?

Ethical considerations in Alzheimer's disease care involve respecting the autonomy and dignity of individuals while ensuring their safety and well-being. Decisions about medical treatments, living arrangements, and end-of-life care should be made with the individual's wishes and values in mind. Balancing the need for independence with the necessity of supervision and intervention requires thoughtful ethical consideration and compassionate communication.

Q35: How do cultural differences impact the perception and management of Alzheimer's disease?

Cultural differences can significantly impact the perception and management of Alzheimer's disease. Beliefs about aging, dementia, and caregiving vary across cultures, influencing how symptoms are recognized and addressed. Cultural norms and values also affect the willingness to seek medical help and the types of support systems available. Culturally sensitive care approaches are essential for effectively supporting individuals with Alzheimer's and their families.

Q36: What is the role of palliative care in Alzheimer's disease?

Palliative care focuses on improving the quality of life for individuals with serious illnesses, including Alzheimer's disease, by addressing their physical, emotional, and spiritual needs. It involves managing symptoms, providing psychosocial support, and assisting with decision-making about care goals. Palliative care can be introduced at any stage of Alzheimer's and is especially important in the advanced stages to ensure comfort and dignity.

Q37: How does Alzheimer's disease affect the healthcare system?

Alzheimer's disease places a significant burden on the healthcare system due to the high cost of long-term care, medical treatments, and support services. The increasing prevalence of Alzheimer's, combined with an aging population, strains resources and underscores the need for effective care models and funding for research and support programs. Addressing these challenges requires coordinated efforts from healthcare providers, policymakers, and communities.

Q38: What is the impact of Alzheimer's disease on public health?

Alzheimer's disease is a major public health issue with significant implications for individuals, families, and society. The growing number of cases demands substantial healthcare resources, long-term care services, and support systems. Public health initiatives focus on raising awareness, promoting early detection and intervention, and supporting research into prevention and treatment. Addressing Alzheimer's disease requires a comprehensive public health approach.

Q39: What were the early reactions from the medical community to Dr. Alzheimer's discovery?

The medical community's initial reaction to Dr. Alois Alzheimer's discovery was relatively subdued. When Dr. Alzheimer presented his findings on Auguste Deter in 1906, his work garnered some interest but did not spark widespread recognition or research. At the time, dementia was largely seen as a normal part of aging, and many physicians did not yet see the need to distinguish between different types of cognitive decline in older adults.

It wasn't until 1910 that Dr. Emil Kraepelin, a prominent psychiatrist and colleague of Alzheimer, formally named the condition "Alzheimer's disease" in his *Textbook of Psychiatry*. By doing so, Kraepelin helped establish it as a distinct condition, separate from other forms of dementia. However, even then, Alzheimer's disease remained a relatively rare and obscure diagnosis for much of the 20th century. It was only as more cases were documented among aging populations that the broader medical community began recognizing it as a major public health issue. Advances in neuroscience and brain imaging in the latter part of the 20th century eventually brought Dr. Alzheimer's work into the spotlight, leading to an explosion of research into the disease.

Q40: How did the understanding of Alzheimer's disease evolve throughout the 20th century?

In the early 20th century, Alzheimer's disease was regarded as a rare form of dementia, primarily affecting younger individuals. After Dr. Alois Alzheimer's initial discovery in 1906, senile dementia was considered a separate, age-related decline that affected older adults, while Alzheimer's was often referred to as presenile dementia due to its association with middle-aged individuals. At the time, dementia was generally understood as an inevitable part of aging, with little differentiation between the various causes of cognitive decline. As a result, Alzheimer's remained an obscure diagnosis, with most physicians viewing dementia as a singular phenomenon rather than a set of distinct diseases.

By the mid-20th century, medical understanding began to evolve as researchers identified specific brain abnormalities—amyloid plaques and neurofibrillary tangles—associated with Alzheimer's disease, which were distinct from those seen in other dementias. This discovery was pivotal in establishing Alzheimer's as a unique disorder with its own pathological profile. Simultaneously, research into other types of dementia, such as vascular dementia (caused by reduced blood flow to the brain) and Lewy body dementia (characterized by abnormal protein deposits), demonstrated that dementia could result from various distinct causes. This broader understanding of dementia helped differentiate Alzheimer's disease, highlighting it as a major neurodegenerative disorder with unique biological markers and progression.

From the 1980s onward, rapid advancements in neuroscience, genetics, and brain imaging provided deeper insights into Alzheimer's disease. Genetic research uncovered specific risk factors, such as the APOE ε4 allele, which increases susceptibility to late-onset Alzheimer's. Brain imaging technologies like PET scans enabled scientists to observe amyloid and tau pathologies in living patients, allowing for earlier and more accurate diagnoses. These developments transformed Alzheimer's from an uncommon disorder to a leading cause of dementia worldwide. The late 20th century also saw increased funding and research into Alzheimer's as a public health priority, distinguishing it from other forms of dementia and solidifying its status as a prevalent disease with significant societal impact.

Q41: What are the key challenges in Alzheimer's disease research?

Alzheimer's disease research faces several key challenges, including understanding the complex biology of the disease, developing effective treatments, and identifying reliable biomarkers for early diagnosis. Additional challenges include securing research funding, addressing ethical considerations in clinical trials, and ensuring diverse study populations. Collaborative efforts and innovative approaches are essential for advancing knowledge and finding solutions to these challenges.

Q42: What are the key areas of focus in Alzheimer's disease advocacy?

Alzheimer's disease advocacy focuses on raising awareness, securing funding for research, enhancing care and support services, and influencing public policy. Advocates work to reduce stigma, educate communities, and promote early detection and intervention. Their efforts also aim to support caregivers and ensure individuals with Alzheimer's receive high-quality, person-centered care. Collaborative efforts among organizations, healthcare providers, and policymakers are crucial for advancing these goals.

–

As we conclude this chapter "Introduction to Alzheimer's Disease," we hope that the insights shared have provided a clearer understanding of the complexities of this condition and its profound impact on individuals, families, and society. Alzheimer's disease presents significant challenges, but with ongoing research and advances in care, there is hope for improved outcomes and quality of life for those affected.

Early diagnosis remains vital in managing Alzheimer's, as it allows for timely interventions and access to supportive services. Equally important is the role of caregivers and support networks in providing compassionate care and fostering well-being throughout the journey with Alzheimer's.

Bibliography

1. ***Alzheimer's Association. (2021).*** *"2021 Alzheimer's disease facts and figures." Alzheimer's & Dementia, 17(3), 327–406.*
2. ***Holtzman, D. M., Morris, J. C., & Goate, A. M. (2011).*** *"Alzheimer's disease: The challenge of the second century." Science Translational Medicine, 3(77), 77sr1.*
3. ***Jack Jr, C. R., Knopman, D. S., Jagust, W. J., Petersen, R. C., Weiner, M. W., Aisen, P. S., Shaw, L. M., Vemuri, P., Wiste, H. J., Weigand, S. D., Lesnick, T. G., Pankratz, V. S., Donohue, M. C., & Trojanowski, J. Q. (2013).*** *"Tracking pathophysiological processes in Alzheimer's disease: An updated hypothetical model of dynamic biomarkers." The Lancet Neurology, 12(2), 207–216.*
4. ***Selkoe, D. J., & Hardy, J. (2016).*** *"The amyloid hypothesis of Alzheimer's disease at 25 years." EMBO Molecular Medicine, 8(6), 595–608.*

5. ***Sperling, R. A., Aisen, P. S., Beckett, L. A., Bennett, D. A., Craft, S., Fagan, A. M., Iwatsubo, T., Jack Jr, C. R., Kaye, J., Montine, T. J., Park, D. C., Reiman, E. M., Rowe, C. C., Siemers, E., Stern, Y., Yaffe, K., Carrillo, M. C., Thies, B., Morrison-Bogorad, M., Wagster, M. V., & Phelps, C. H. (2011).*** *"Toward defining the preclinical stages of Alzheimer's disease: Recommendations from the National Institute on Aging-Alzheimer's Association workgroups on diagnostic guidelines for Alzheimer's disease." Alzheimer's & Dementia, 7(3), 280–292.*

6. ***Zhang H, Jiang X, Ma L, Wei W, Li Z, Chang S, Wen J, Sun J and Li H*** *(2022) Role of Aβ in Alzheimer's-related synaptic dysfunction. Front. Cell Dev. Biol. 10:964075. doi: 10.3389/fcell.2022.964075*

CHAPTER 2 Causes and Risk Factors

Welcome to the chapter on "Causes and Risk Factors" of Alzheimer's disease, where we explore the diverse factors contributing to the development of this complex neurodegenerative condition. Alzheimer's disease affects millions worldwide, presenting ongoing challenges to our understanding of its origins and risk factors. By examining the interplay of genetic, environmental, and lifestyle influences, we aim to shed light on the underlying mechanisms that drive Alzheimer's disease onset and progression.

This Q&A chapter will explore the known causes and established risk factors associated with Alzheimer's disease. We will discuss genetic mutations linked to familial forms of the disease, analyze the role of age and demographic factors, and examine lifestyle choices that may influence susceptibility. Our goal is to provide evidence-based insights that empower individuals, families, and healthcare professionals to make informed decisions regarding prevention, early detection, and personalized management strategies.

–

Q1: What causes Alzheimer's disease?

Alzheimer's disease is not a straightforward condition with a single cause. Rather, it is a complex neurodegenerative disorder that arises from a series of interconnected events over time, gradually affecting the brain. The disease is believed to result from a combination of genetic, environmental, lifestyle, and other factors that are still being identified. Key pathological features include the accumulation of amyloid plaques and tau tangles in the brain, which disrupt neural communication and lead to cell death. Genetic factors, such as mutations in the APP, PSEN1, and PSEN2 genes, can increase the risk of developing early-onset Alzheimer's. Additionally, the APOE ε4 allele is a significant risk factor for late-onset Alzheimer's. The complex interplay between genetic and non genetic factors involved in the development of the disease makes it challenging for researchers to identify a precise treatment.

After a diagnosis of Alzheimer's disease, life expectancy is typically reduced by about one-third. On average, individuals with Alzheimer's live for 8 to 10 years following diagnosis, though the disease can persist for up to 20 years. The progression of Alzheimer's and the severity of symptoms can vary significantly from person to person.

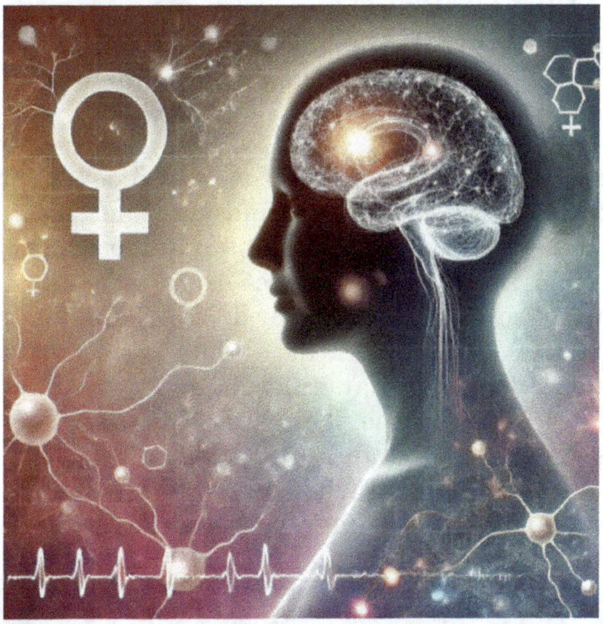

▲ **FIGURE 2.1.**
Hormonal effects on AD (women).

Q2: What are the genetic causes and risk factors for Alzheimer's disease?

Alzheimer's disease is caused by a combination of genetic and environmental factors, and it doesn't have a single, simple cause. In some families, a small number of individuals develop

Alzheimer's at a younger age, before 60, due to changes in specific genes located on chromosomes 21, 14, and 1. These genetic changes are passed down through families in a pattern called autosomal-dominant inheritance. This means that if a parent has this form of Alzheimer's, there is a 50% chance that their children, siblings, or other close relatives will also develop the disease. This early-onset form of Alzheimer's is known as *familial Alzheimer's* and accounts for only about 5% of all Alzheimer's cases.

However, most individuals with Alzheimer's develop the disease after the age of 65, which is referred to as *late-onset Alzheimer's*. The primary gene linked to late-onset Alzheimer's is the APOE gene. This gene has three versions, or *alleles*: 2, 3, and 4. Having the APOE-4 allele increases the risk of developing Alzheimer's, but not everyone with this allele will develop the disease. Conversely, the APOE-2 allele can offer some protection against Alzheimer's, making it less likely to develop. While the APOE gene doesn't directly cause Alzheimer's, it influences the age at which symptoms may appear.

In addition to APOE, other genes may also contribute to late-onset Alzheimer's, such as those found on chromosome 12, though scientists are still working to understand their exact role. These findings highlight the complexity of Alzheimer's disease. While genetics play a significant role, they don't account for everything. For example, even identical twins—who share the same genes—can develop Alzheimer's at different ages, or only one twin might develop the disease. This suggests that environmental factors, such as lifestyle and health habits, also influence the likelihood and timing of Alzheimer's onset. Therefore, genetic and environmental factors likely work together to determine the course of the disease.

Q3: What is the relationship between sleep and Alzheimer's disease?

The relationship between sleep and Alzheimer's disease is complex, with research indicating that poor sleep can contribute to both the development and progression of Alzheimer's. Conversely, the disease itself can disrupt sleep patterns. One key connection is that sleep aids the brain in clearing harmful waste products, including *beta-amyloid*, a protein that accumulates in the brains of people with Alzheimer's. During deep sleep, the brain's glymphatic system becomes more active, facilitating the removal of these toxins. When sleep is disrupted or insufficient, this cleaning process is impaired, leading to a buildup of beta-amyloid, which may increase the risk of Alzheimer's.

Moreover, people with Alzheimer's often face sleep disturbances, such as difficulty falling asleep, frequent nighttime awakenings, and early morning waking. These sleep issues can exacerbate cognitive (thinking and memory) problems. Poor sleep is also linked to other factors that elevate the risk of Alzheimer's, such as inflammation, memory issues, and stress. Therefore, maintaining healthy sleep habits may help prevent Alzheimer's or slow its progression, particularly in individuals already showing signs of the disease.

Q4: How do lifestyle factors influence the risk of Alzheimer's disease?

Lifestyle factors play a significant role in influencing the risk of developing Alzheimer's disease, with growing evidence suggesting that certain behaviors can either increase or decrease risk. Physical activity, diet, social engagement, cognitive stimulation, sleep quality, and cardiovascular health management have all been shown to have a considerable impact on brain health.

Regular physical activity has consistently been associated with a lower risk of Alzheimer's. Exercise increases blood flow to the brain, supports the growth of new neurons, and helps reduce inflammation. Additionally, engaging in a balanced diet—particularly the Mediterranean or DASH diets, which are rich in vegetables, fruits, whole grains, and healthy fats—has been linked to a reduced risk of Alzheimer's, as these diets support brain health and reduce oxidative stress.

Social connections and cognitive engagement through mentally stimulating activities, such as learning new skills, solving puzzles, or reading, act as protective factors. These activities help maintain cognitive reserves, providing a buffer against age-related decline. Quality sleep is also crucial; poor sleep and sleep disorders, such as sleep apnea, have been associated with an increased risk of Alzheimer's, possibly due to the impaired clearance of harmful proteins from the brain during sleep.

Managing cardiovascular health—by controlling blood pressure, cholesterol, and blood sugar—can also reduce the risk of Alzheimer's. Conditions such as hypertension, obesity, and diabetes are associated with an increased risk of Alzheimer's because they impact blood flow and increase inflammation in the brain.

Q5: Is age a risk factor for Alzheimer's disease?

Age is the most significant risk factor for Alzheimer's disease, with the likelihood of developing the condition increasing substantially after age 65 and roughly doubling every 5 years thereafter. While Alzheimer's can occur in younger individuals, particularly in those with a strong genetic predisposition, the vast majority of cases occur in older adults. The percentage of people with Alzheimer's rises sharply with age: Approximately 3% of those aged 65–74, 17% of those aged 75–84, and 32% of individuals aged 85 or older have Alzheimer's dementia. However, it's important to note that Alzheimer's is not a normal part of aging; advancing age alone does not cause the disease.

Q6: Is gender a risk factor for Alzheimer's disease?

Gender plays a role in Alzheimer's disease risk, with women being more likely to develop the condition than men. This increased risk is partly due to women generally living longer than men, but there may also be biological factors that contribute to this difference. Ongoing research aims to better understand the specific reasons behind the gender disparity in Alzheimer's prevalence.

Q7: Is race a risk factor for Alzheimer's disease?

Alzheimer's disease affects different racial and ethnic groups in varying ways. Research suggests that African Americans and Hispanic Americans may have a higher risk of developing Alzheimer's compared to Caucasians. However, more studies are needed to fully understand why these disparities exist. These differences may be influenced by a combination of genetic, environmental, and socioeconomic factors. Additionally, as non-Caucasian populations live longer, they will constitute a larger portion of the aging population. By 2050, it is projected that the percentage of people over 85 who are non-Caucasian will increase from 16% to 34%, underscoring the growing need to address Alzheimer's in these communities.

Q8: How does education impact the risk of Alzheimer's disease?

People with more years of formal education are at a lower risk of developing Alzheimer's and other dementias compared to those with less education. One possible reason for this is the concept of *cognitive reserve*. Cognitive reserve refers to the brain's ability to adapt and compensate for damage, allowing individuals to maintain cognitive functions despite age-related changes or Alzheimer's pathology. This enables the brain to continue functioning well even as it undergoes dementia-related changes, such as the buildup of specific proteins. Cognitive reserve isn't only developed through education—mentally challenging work and hobbies also contribute.

Individuals with fewer years of education are often in lower-income situations, which can lead to challenges such as poor nutrition and limited access to healthcare, including treatments for heart health. They may also be less physically active and face a higher risk of conditions like diabetes and heart disease, which are linked to an increased likelihood of developing dementia.

Q9: What is the role of cardiovascular health in Alzheimer's disease?

Brain health is closely linked to heart and blood vessel health, as the brain requires 20% of the body's oxygen and energy. A strong heart and healthy blood vessels ensure a steady flow of oxygen- and nutrient-rich blood to the brain. Several factors that increase the risk of heart disease are also associated with a higher risk of dementia. These include smoking and diabetes. Evidence suggests that poor glucose processing, which can lead to diabetes, may also increase the risk of dementia. Additionally, risk factors such as obesity, high blood pressure, and high cholesterol—especially if they develop in midlife—are linked to an increased risk of dementia. However, these same factors might be associated with a lower risk when they appear later in life. More research is needed to understand how the timing of these risks affects dementia.

Studies indicate that factors promoting heart health may also help protect the brain, potentially lowering the risk of Alzheimer's or other dementias. For example, regular physical activity benefits both heart and brain health, though researchers are still exploring the most effective types, frequencies, and duration of exercise. A heart-healthy diet that includes fruits, vegetables, whole grains, fish, chicken, nuts, and legumes while limiting saturated fats, red meat, and sugar may also help reduce dementia risk.

Q10: How do lifestyle, diet, and conditions like diabetes affect the risk of Alzheimer's disease?

Lifestyle factors such as Type 2 diabetes (T2D), heart disease, midlife obesity, and insulin resistance are all linked to an increased risk of Alzheimer's disease. These conditions disrupt normal body functions and are often associated with chronic inflammation and oxidative stress, where an imbalance of free radicals and antioxidants leads to cellular damage, including in the brain. T2D, in particular, has been shown to contribute to cognitive decline, with individuals with diabetes experiencing worse cognitive function and faster mental deterioration compared to those without the condition. The metabolic issues tied to T2D, including insulin resistance and high blood sugar, can exacerbate brain degeneration and may also increase the risk of stroke and heart disease, further contributing to cognitive decline.

Diet plays a significant role in reducing Alzheimer's risk, with evidence suggesting that foods rich in polyphenols—such as green vegetables, berries, herbs, and spices—can help protect brain health by acting as antioxidants. Diets like the Mediterranean and Okinawan diets, which emphasize antioxidant-rich foods and balance healthy fats (increasing omega-3s and reducing omega-6s and saturated fats), have been linked to longer life and better cognitive health. Additionally, the gut– brain connection, influenced by the digestion of carbohydrates and proteins and the role of gut bacteria, also appears to impact brain function and Alzheimer's risk.

Q11: How does chronic stress impact the risk of Alzheimer's disease?

The *vicious cycle of stress* refers to how stress contributes to the development of disease, and how the disease itself exacerbates stress, accelerating the illness's progression and leading to mental health complications. In the context of Alzheimer's disease, stress plays a critical role in both its development and worsening. Stress triggers the release of stress hormones, such as cortisol, which, in moderate amounts, helps the body respond to challenges. However, chronic or prolonged stress results in elevated cortisol levels, which can damage the brain.

In Alzheimer's, high cortisol levels harm the hippocampus, a region essential for memory and learning, making it more difficult for individuals to retain information and manage stress. This damage contributes to cognitive decline and exacerbates Alzheimer's symptoms. At the same time, the confusion, memory loss, and problem-solving difficulties caused by Alzheimer's heighten feelings of anxiety and frustration, leading to increased stress. This heightened stress can further raise cortisol levels, perpetuating the cycle.

The cycle of stress and cognitive decline also makes it more challenging for individuals with Alzheimer's to cope with everyday tasks, intensifying their stress and worsening symptoms. In this way, anxiety and Alzheimer's feed into each other, creating a destructive loop.

Q12: How do social connections affect the risk of Alzheimer's disease?

Engaging in social activities and maintaining close relationships with family, friends, and the community can help preserve cognitive function as we age. Social interactions stimulate mental activity, which strengthens neural networks—the pathways connecting brain cells—and helps build cognitive reserve. Research has shown that people who feel socially connected and supported are at a lower risk of cognitive decline compared to those who experience isolation or loneliness.

In contrast, social isolation and loneliness can increase stress levels and contribute to depression, both of which are associated with a higher risk of Alzheimer's. Chronic stress resulting from loneliness may lead to harmful inflammation, weakening the brain's resistance to neurological diseases. Therefore, fostering strong social connections may play a protective role against Alzheimer's, helping to maintain mental sharpness and overall brain health. Activities that involve social interaction, such as group hobbies, volunteering, and spending time with family and friends, can help protect against cognitive decline.

▲ **FIGURE 2.3.**
An artistic representation in the mind of an Alzheimer's patient.

Q13: How does smoking affect the risk of Alzheimer's disease?

Smoking significantly increases the risk of Alzheimer's disease and other forms of dementia. Studies have shown that smoking damages blood vessels, reducing blood flow to the brain and increasing the risk of vascular issues such as strokes, which are linked to dementia. Smoking also exposes the brain to high levels of oxidative stress and inflammation, which can damage neurons and contribute to the formation of plaques and tangles associated with Alzheimer's disease.

Additionally, chemicals in tobacco smoke can increase the likelihood of developing health conditions such as high blood pressure, high cholesterol, and type 2 diabetes—all of which are known risk factors for Alzheimer's. Quitting smoking and avoiding second-hand exposure can help reduce these risks and protect brain health. Stopping smoking allows the body to begin repairing itself and can slow or prevent further damage to brain cells.

Q14: How does alcohol consumption affect the risk of Alzheimer's disease?

Alcohol consumption can have a complex impact on the risk of Alzheimer's disease, with effects varying depending on the amount consumed. Studies suggest that heavy or long-term alcohol use may increase the risk of Alzheimer's due to brain cell damage, inflammation, and reduced brain volume. Alcohol abuse is also linked to conditions like liver disease, cardiovascular disease, and diabetes, which further elevate dementia risk.

Heavy drinking can impair the brain's communication networks and reduce the production of new brain cells, negatively affecting memory and cognitive functions. It also contributes to oxidative stress and disrupts blood flow to the brain, both of which are associated with the types of damage seen in Alzheimer's disease.

On the other hand, some research suggests that light to moderate alcohol consumption—such as one drink per day—may be linked to a lower risk of dementia, potentially due to improved cardiovascular health, which benefits brain health as well. However, these findings remain debated, as the protective effects may vary depending on individual health factors, and even moderate drinking may carry risks for some people. In general, while moderate alcohol intake might not significantly increase Alzheimer's risk for some individuals, heavy and long-term drinking is associated with a higher likelihood of developing the disease.

Q15: How does physical activity influence the risk of Alzheimer's disease?

Maintaining a healthy body supports a healthy mind, and exercise plays a crucial role in weight management, circulation, and brain function. Regular physical activity helps protect and stimulate brain performance, especially in individuals showing early signs of age-related brain diseases. This benefit may stem from exercise's ability to boost oxygen flow to the brain, enhance brain signaling speed, and increase the production of neurotrophic factors—proteins that support brain cell growth and health. Activities such as walking, swimming, and strength training can be tailored to fit various fitness levels and abilities, making exercise an essential part of Alzheimer's disease management.

Q16: How does cognitive engagement impact the risk of Alzheimer's disease?

Engaging in mentally stimulating activities can reduce the risk of developing Alzheimer's disease by building cognitive reserve. Activities that challenge the mind—such as reading, learning new skills, solving puzzles, playing games, and problem-solving—are believed to strengthen the brain's neural networks, helping to preserve mental sharpness.

Research indicates that those who regularly engage in mentally stimulating activities have a lower risk of Alzheimer's and other forms of dementia. This may be due to stronger neural connections and increased production of chemicals that support memory and learning. Social activities that involve cognitive engagement, such as group games or discussions, can enhance these benefits by combining mental stimulation with social interaction, another key factor in brain health.

While cognitive engagement does not guarantee protection from Alzheimer's, it can help delay the onset of symptoms or reduce their severity by building resilience in brain function. This underscores the importance of staying mentally active throughout life as part of a brain-healthy lifestyle.

Q17: What is the impact of head injuries on Alzheimer's disease risk?

Head injuries, particularly traumatic brain injuries (TBIs), are linked to an increased risk of Alzheimer's disease and other dementias. A *TBI* refers to any impact to the head that disrupts normal brain function, ranging from mild (like a concussion) to severe. The risk of dementia is generally higher for individuals who have experienced multiple TBIs or moderate to severe injuries, with the risk increasing in proportion to the severity and frequency of injuries.

After a TBI, the brain may experience inflammation and other changes that damage neurons, the cells responsible for transmitting information. This damage can lead to the buildup of abnormal proteins, such as beta-amyloid and tau, which are hallmarks of Alzheimer's disease. Some research suggests that TBIs may activate pathways in the brain that contribute to Alzheimer's progression, potentially accelerating symptoms in those already at risk.

Although not everyone with a history of head injury will develop dementia, the increased risk emphasizes the importance of protecting the head from injury, particularly in high-impact activities or sports. Proper use of helmets, seat belts, and other safety measures can reduce the likelihood of head injury and, in turn, may help lower the risk of Alzheimer's later in life.

Q18: How do infections and inflammation influence the risk of Alzheimer's disease?

Chronic inflammation, particularly in the brain, has been linked to an increased risk of Alzheimer's and can worsen its symptoms. Microglia, the brain's immune cells, typically help clear damaged cells and harmful substances, including beta-amyloid plaques, which are a hallmark of Alzheimer's disease. However, when inflammation becomes prolonged or excessive, microglia can become overactive, damaging healthy brain tissue and contributing to the buildup of toxic proteins like beta-amyloid and tau. This creates a cycle of neurodegeneration, where inflammation accelerates brain damage and cognitive decline.

Infections, particularly those affecting the brain, such as viral or bacterial infections, can also increase the risk of Alzheimer's. These infections trigger immune responses in the brain, causing inflammation that can damage neurons and accelerate Alzheimer's progression. Some studies suggest that specific infections, like the herpes simplex virus (HSV), may be linked to an increased risk of Alzheimer's in individuals already predisposed to the disease.

Furthermore, conditions that cause widespread inflammation in the body, such as diabetes, cardiovascular disease, or obesity, may also increase Alzheimer's risk by promoting inflammation that extends to the brain. Managing inflammation through a healthy diet, regular exercise, and proper treatment of infections or inflammatory conditions can help reduce the risk of Alzheimer's and slow its progression.

Q19: What role does air pollution play in the risk for Alzheimer's disease?

Air pollution is increasingly recognized as a potential risk factor for Alzheimer's disease. Long-term exposure to air pollutants, including fine particulate matter (PM2.5), nitrogen dioxide (NO2), and ground-level ozone, has been linked to cognitive decline and an increased risk of developing dementia, including

Alzheimer's disease. These pollutants can reach the brain directly through the bloodstream or by causing inflammation in the respiratory system, which can then affect the brain.

Studies suggest that air pollution may contribute to Alzheimer's through several mechanisms. For example, pollutants can trigger inflammation in the brain, which damages neurons and promotes the buildup of toxic proteins, such as beta-amyloid and tau, that are characteristic of Alzheimer's. Chronic inflammation can lead to neuronal damage, accelerating cognitive decline.

Additionally, air pollution impacts the cardiovascular system, which in turn affects brain health. Poor heart health can reduce blood flow to the brain, depriving it of the oxygen and nutrients necessary for proper function. This reduced blood flow may contribute to neurodegenerative processes, making the brain more susceptible to conditions like Alzheimer's. There is also evidence suggesting that air pollution may influence brain development and aging. Studies have found that children and older adults living in areas with high pollution levels may experience changes in brain structure and function, which could increase their risk of developing Alzheimer's later in life.

While more research is needed to fully understand the relationship between air pollution and Alzheimer's, existing studies emphasize the importance of reducing exposure to environmental pollutants to protect brain health.

Q20: How do hormonal changes impact Alzheimer's disease risk?

Two-thirds of all dementia cases occur in women, making it essential to understand how hormonal changes, especially those related to estrogen, contribute to brain health. Hormones play a crucial role in various factors that influence brain function, including blood vessel health (vascular risk), inflammation, and the development of diseases like dementia. A key area of focus is how the decline in estrogen levels over several decades—often starting around midlife and continuing into old age—may contribute to the onset of dementia symptoms.

Understanding the impact of estrogen loss on the brain during this time is critical for identifying potential treatments to prevent or slow dementia. Estrogen is thought to help protect the brain, and as levels decrease, the risk of neurodegenerative diseases, such as Alzheimer's, may rise. One important consideration is the timing and type of hormone therapy (HT) that could be most beneficial. Research is needed to determine the optimal timing for starting hormone therapy and which women would benefit most. Further studies are required to deepen our understanding of how women's brains change during menopause and aging. Expanding this research could help determine whether hormone therapy may reduce the risk of dementia or improve brain function in older women.

Q21: How does marijuana usage affect the risk of Alzheimer's disease?

Long-term, heavy marijuana use, especially with high THC content, may increase the risk of cognitive decline, including memory impairment— a key symptom of Alzheimer's disease. THC interacts with the brain's endocannabinoid system, affecting memory, learning, and cognitive processing, particularly in the hippocampus, a region essential for memory that is often among the first to be impacted in Alzheimer's. Studies have shown that chronic marijuana use can alter the structure and function of the hippocampus over time, potentially accelerating age-related memory problems and cognitive decline. Given that Alzheimer's already affects memory and cognition, heavy marijuana use could exacerbate these issues, potentially increasing the risk of developing Alzheimer's or related dementias later in life.

Another concern with marijuana use is its potential impact on brain blood flow and vascular health. Some studies suggest that heavy marijuana use may reduce blood flow to specific areas of the brain, particularly the hippocampus. This is worrying because poor vascular health and reduced blood flow are linked to an increased risk of Alzheimer's and other dementias. Furthermore, prolonged use of high-THC marijuana may interfere with brain development if started at a young age. Research shows that early marijuana use is associated with more significant long-term cognitive effects. Together, these findings suggest that long-term, heavy marijuana use—particularly in vulnerable individuals—may increase the risk of Alzheimer's by worsening memory issues, diminishing cognitive resilience, and negatively impacting brain vascular health.

–

As we conclude this chapter on the " Causes and Risk Factors of Alzheimer's Disease, " the factors contributing to this complex neurological condition are multifaceted. While the exact origins of Alzheimer's remain unclear, ongoing research is shedding light on its etiology and progression.

Understanding the interplay of genetic predispositions, environmental exposures, and lifestyle choices is crucial in developing effective strategies for prevention and early intervention. By focusing on modifiable risk factors such as cardiovascular health, physical activity, and cognitive engagement, individuals may be able to reduce their risk of developing Alzheimer's disease and support long-term brain health.

Bibliography

1. ***Paulavicius, A. M., Mizzaci, C. C., Tavares, D. R. B., Rocha, A. P., Civile, V. T., Schutz, R. R., Pereira Nunes Pinto, A. C., & Trevisani, V. F. M.*** *"Bilingualism for delaying the onset of Alzheimer's disease: A systematic review and meta-analysis." European Geriatric Medicine, 11(4), 651–658. https://doi.org/10.1007/s41999-020-00326-x*
2. ***Alzheimer's Association.*** *(2019). "2019 Alzheimer's disease facts and figures." Alzheimer's & Dementia : The Journal of the Alzheimer's Association , 15(3), 321–387. https://doi.org/10.1016/j.jalz.2019.01.010*
3. ***Bertram, L., & Tanzi, R. E.*** *(2008). " Thirty years of Alzheimer's disease genetics: The implications of systematic meta-analyses. " Nature Reviews Neuroscience, 9(10), 768– 778.*
4. ***Brookmeyer, R., Johnson, E., Ziegler-Graham, K., & Arrighi, H. M.*** *(2007). " Forecasting the global burden of Alzheimer's disease. " Alzheimer's & Dementia, 3(3), 186– 191.*
5. ***Coulter, K., & Phillips, N. A.*** *(2024). "Bilinguals show evidence of brain maintenance in Alzheimer's disease." Bilingualism: Language and Cognition , 27(5), 1029–1038. https://doi.org/10.1017/S1366728924000221*
6. ***Ferretti, M. T., Dimech, A. S., & Chadha, A. S.*** *(Eds.). (2021). Sex and gender differences in Alzheimer's disease : The women's brain project. Academic Press.*
7. ***Gatz, M., Reynolds, C. A., Fratiglioni, L., Johansson, B., Mortimer, J. A., Berg, S., Fiske, A., & Pedersen, N. L.*** *(2006). " Role of genes and environments for explaining Alzheimer disease. " Archives of General Psychiatry, 63(2), 168– 174.*
8. ***Gruetzner, H.*** *(2001). Alzheimer's : A caregiver's guide and sourcebook (3rd ed.). Wiley.*
9. ***Justice, N. J.*** *(2018). "The relationship between stress and Alzheimer's disease." Neurobiology of Stress, 8, 127–133. https://doi.org/10.1016/j.ynstr.2018.04.002*
10. ***Martins, R. N., Brennan, C. S., Fernando, B., Brennan, M. A., & Fuller, S. J.*** *(2019). Neurodegeneration and Alzheimer's Disease : The Role of Diabetes, Genetics, Hormones, and Lifestyle (1st ed.). Wiley & Sons . https://doi.org/10.1002/9781119356752*

11. ***Norton, S., Matthews, F. E., Barnes, D. E., Yaffe, K., & Brayne, C.*** *(2014). "Potential for primary prevention of Alzheimer's disease: An analysis of population-based data. " The Lancet Neurology, 13(8), 788– 794.*

12. ***Panza, F., Lozupone, M., Seripa, D., & Imbimbo, B. P.*** *(2019). " Diet and Alzheimer's disease risk factors or prevention: The current evidence. " Expert Review of Neurotherapeutics, 19(7), 677– 708.*

13. ***Paulavicius, A. M., Mizzaci, C. C., Tavares, D. R. B., Rocha, A. P., Civile, V. T., Schultz, R. R., Pinto, A. C. P. N., & Trevisani, V. F. M.*** *(2020). "Bilingualism for delaying the onset of Alzheimer's disease: a systematic review and meta-analysis." European Geriatric Medicine, 11(4), 651–658.*

Symptoms and Diagnosis

CHAPTER 3

This chapter is a key section of this comprehensive guide, focusing on the early indicators and diagnostic approaches related to Alzheimer's disease. Recognizing the symptoms of Alzheimer's at an early stage is essential for effective intervention, as early detection can significantly influence the management and progression of the disease. This chapter examines the cognitive, behavioral, and physical changes commonly seen in individuals with Alzheimer's, distinguishing these symptoms from normal age-related changes. Additionally, it provides an overview of the diagnostic process, including medical evaluations, cognitive assessments, and advanced imaging techniques, offering readers a thorough understanding of how Alzheimer's is identified and diagnosed.

–

Q1: What are the signs and symptoms of Alzheimer's disease?

Alzheimer's disease is characterized by a progressive decline in intellectual abilities, significantly impacting an individual's capacity to carry out daily activities. One of the earliest signs is a noticeable decline in cognitive functions, including memory, reasoning, and problem-solving. Once easily managed, routine tasks become increasingly challenging, with individuals struggling to perform basic activities such as cooking, managing finances, or driving.

Memory problems are central to Alzheimer's, with individuals often forgetting recent events or conversations, while memories of the distant past tend to persist longer. This contrast can create confusion and a sense of disconnection from their environment. Impaired judgment is another common symptom, as individuals may fail to recognize dangerous situations, such as leaving appliances on or engaging in risky behaviors. This loss of judgment extends to decision-making tasks, like managing money or determining when help is needed.

As the disease progresses, individuals may have trouble differentiating between objects, faces, and concepts, leading to confusion and a reduced ability to recognize familiar people and everyday items. Aphasia, a condition affecting speech and language comprehension, can make it difficult for individuals to find the right words or understand conversations. Additionally, purposeful movements become challenging, with individuals struggling to perform tasks requiring coordination, such as dressing or eating. Apraxia, the inability to carry out planned movements, often leads to a loss of independence in performing simple actions.

Recognition of familiar people and objects may also decline, leaving individuals unable to identify loved ones or everyday items. Mathematical skills and the ability to understand basic concepts, such as time and numbers, diminish over time, leading to difficulties with budgeting or reading a clock. Writing abilities also deteriorate, requiring assistance to write legible sentences or maintain clear communication.

Personality changes, such as mood swings, anxiety, or a loss of interest in previously enjoyed activities, often emerge early in the

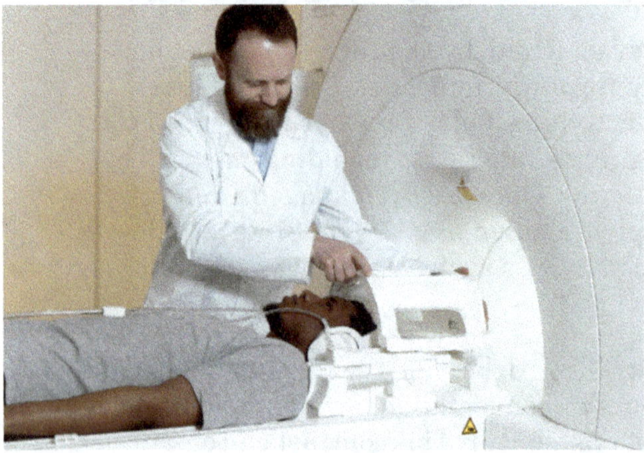

▲ **FIGURE 3.1.**
Photo of doctor examining patient.

disease. These shifts can be distressing for both the individual with Alzheimer's and their loved ones, as the person may appear to drift away from their former self.

Q2: How is Alzheimer's disease diagnosed in a person?

Alzheimer's disease is diagnosed through a combination of medical history, physical exams, cognitive assessments, and brain imaging. Healthcare professionals such as neurologists (specialists in brain conditions) or geriatricians (specialists in older adults) will review the individual's symptoms, medical history, and medication use. They may also interview a close family member or friend to gather additional information about any changes in memory or behavior. A physical exam and several cognitive tests are conducted to evaluate cognitive skills, such as memory, problem-solving, attention, and language, as well as any behavioral changes and their impact on daily life. These assessments help doctors determine the severity of the condition and rule out other health problems that could be contributing to the symptoms.

Cognitive and neuropsychological tests measure memory, problem-solving skills, attention, and language abilities through tasks like word recall or puzzle-solving. A physical and neurological exam is also performed to exclude conditions that may cause similar symptoms, including checks of reflexes, coordination, and muscle tone. Brain imaging techniques, such as MRI or CT scans, can reveal brain shrinkage or other abnormalities, while PET scans may detect amyloid plaques, which are a hallmark of Alzheimer's. Blood tests are used to rule

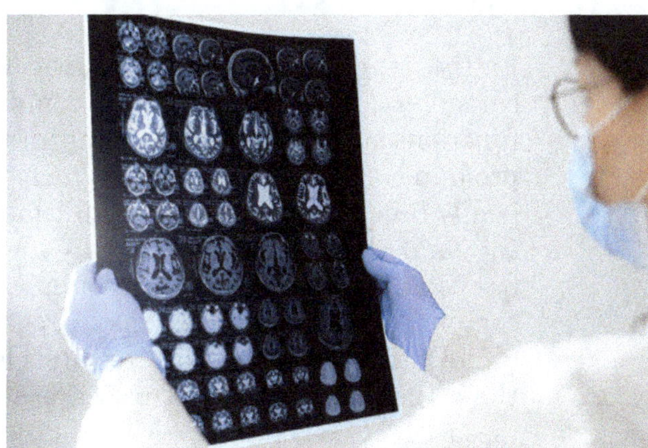

▲ **FIGURE 3.2.**
A doctor holding an MRI result of the brain.

out other causes of cognitive issues, such as vitamin deficiencies or thyroid problems. In some cases, genetic testing may be conducted, particularly in cases of early-onset Alzheimer's, to check for mutations on chromosomes 1, 14, or 21.

Since diagnosing Alzheimer's involves ruling out other potential causes of cognitive decline, the process can be complex and may require a combination of these tests to confirm the diagnosis.

Q3: How does Alzheimer's disease progress over time?

Alzheimer's disease progresses gradually over time, with symptoms worsening in stages. However, the rate of decline can vary from person to person. In the early stage, individuals may experience mild memory loss and confusion, often forgetting recent events or conversations and misplacing items. While daily activities may not be significantly impacted, tasks that require more complex thinking, such as managing finances or organizing, become more difficult. Mood changes, including increased irritability or anxiety, are also common during this stage.

In the middle stage, memory loss becomes more severe, with individuals forgetting personal details and struggling to recognize family members. Cognitive abilities continue to decline, making daily tasks like dressing or cooking increasingly challenging. Communication difficulties intensify, and individuals may become confused about time, place, and events. Behavioral changes such as wandering, aggression, or paranoia are frequent, and the person may require more assistance with daily activities.

In the late stage, individuals become entirely dependent on caregivers, losing the ability to communicate and perform basic functions such as walking, sitting, or swallowing. Memory loss is profound, and they may not recognize loved ones. Motor skills and bodily functions decline, and full-time care becomes essential. In the final stages, individuals may lose the ability to speak, eat, or even move, becoming more vulnerable to infections and other complications. As the disease progresses, complications such as infections or organ failure often lead to death, highlighting the critical need for full-time care and support throughout the disease.

Q4: How often does early-stage Alzheimer's disease go undetected?

The subtle symptoms of early-stage Alzheimer's disease and Mild Cognitive Impairment (MCI) can make detection challenging, especially during brief primary care visits where individuals may appear to function appropriately. Consequently, these conditions are frequently overlooked. Studies focusing on MCI in primary care settings within the United States Medicare population indicate that, on average, only about 8% of expected MCI cases are diagnosed. Other research based on self-reported diagnoses found that only 11% to 15% of individuals with MCI were aware of their diagnosis. Similarly, a German study showed that primary care physicians correctly identified only 11– 12% of established MCI cases. More broadly, it's estimated that while about 20% of primary care patients aged 65 and older are expected to have MCI or dementia, only 8% have received such a diagnosis. In the UK, it has been noted that only about two-thirds of dementia cases are recorded in primary care records. This underdiagnosis, particularly in the earliest symptomatic stages like MCI, means many individuals do not receive timely support or access to interventions. Unaddressed cognitive problems can lead to potentially harmful consequences for patients.

Establishing social support systems early in the disease process is essential for managing care, nutrition, safety, and planning for financial and legal matters. Individuals with Alzheimer's disease are at higher risk for accidents, including while driving, and may not recognize or report other health issues. Early diagnosis enables patients and families to make critical decisions about transportation, living arrangements, and care needs, while the patient is still able to manage daily activities independently.

Q5: What cognitive functions are affected by Alzheimer's disease?

Alzheimer's disease affects multiple cognitive functions, typically starting with memory. Individuals may initially struggle to remember recent events or new information, while older memories often remain more intact. This memory decline is usually followed by a gradual deterioration in executive functioning, which includes essential skills like planning, organizing, and managing tasks. As a result, individuals may become more disorganized, forgetful, or have difficulty completing multi step tasks.

Language abilities are also significantly impacted; people with Alzheimer's may initially have trouble finding words or may resort to vague expressions. As the disease progresses, forming and understanding sentences becomes increasingly challenging, making conversations difficult. Visuospatial skills are affected as well, leading to trouble recognizing familiar objects and faces or navigating well-known places, which can cause balance issues, clumsiness, and even disorientation.

Finally, Alzheimer's disease impairs judgment and comprehension, making it harder for individuals to assess situations accurately or follow complex ideas. Together, these cognitive declines reveal the profound impact Alzheimer's has on various cognitive functions. Memory loss typically leads the progression, followed by declines in other essential skills, ultimately affecting the individual's ability to live independently.

Q6: What are the common behavioral and psychological symptoms of Alzheimer's disease?

Alzheimer's disease can bring about a range of mental and behavioral changes, including anxiety, apathy, depression, hallucinations, and delusions. Individuals may also exhibit behavioral changes such as aggression, restlessness, sleep disturbances, agitation, wandering, and socially inappropriate actions. Although these symptoms are not part of the official diagnostic criteria for Alzheimer's, they are important because they can influence the disease's progression. For example, sleep disturbances, anxiety, and apathy may accelerate the transition from mild cognitive impairment to full-blown dementia.

Furthermore, hallucinations and apathy are linked to greater functional decline and worsening of the disease. Over time, emotional symptoms like depression and anxiety may lessen, but apathy tends to increase. In some cases, individuals may become more irritable or exhibit socially inappropriate behaviors. These emotional and behavioral changes can vary. While Alzheimer's disease itself contributes to these symptoms, it's also important to recognize that they can be triggered or exacerbated by other factors. Such factors can include unmet needs (like hunger or thirst), pain, infections (such as urinary tract infections), medication side effects, constipation, dehydration, or environmental aspects like excessive noise, inadequate lighting, or changes in routine. Investigating and addressing these potential underlying causes is crucial before attributing the behaviors solely to the progression of Alzheimer's. Psychotic symptoms such as hallucinations and delusions are typically more common in the later stages of Alzheimer's but can also appear earlier in other forms of dementia, such as Lewy body dementia.

Q7: How does Alzheimer's disease affect the body?

Dementia is often linked to a range of physical issues that can exacerbate health risks. Some of these physical problems stem from conditions that increase the likelihood of dementia, such as stroke or heart disease, while others develop as dementia progresses. One common issue is difficulty swallowing, which can lead to aspiration pneumonia. People with dementia are hospitalized more frequently for conditions like circulatory problems, infections, and respiratory issues compared to those without dementia.

A significant physical concern associated with cognitive decline is mobility impairment. Although motor and coordination difficulties typically emerge in the later stages of Alzheimer's, balance problems can manifest even in the early stages, such as with mild cognitive impairment (MCI), which often precedes dementia. As mobility becomes more challenging, the risk of falling increases for individuals with dementia. Contributing factors include leg weakness, balance issues, vision problems, low blood pressure, cognitive decline, and medication use. Some fall risk factors are specific to dementia, such as a history of falls, the use of heart-related medications, dizziness upon standing, lack of physical activity, and cognitive impairments affecting executive function, such as difficulty with planning and organizing.

Q8: How does Alzheimer's disease impact memory?

Memory problems are among the earliest and most noticeable signs of Alzheimer's disease. While some memory loss is a normal part of aging, Alzheimer's causes more severe and disruptive memory issues. Forgetting things like keys or appointments can happen to anyone, but in Alzheimer's, memory loss significantly impacts daily life. Individuals with Alzheimer's often struggle to remember recent events or information, such as something they heard just moments ago, and may have difficulty keeping track of bills, appointments, or routines. New information becomes increasingly difficult to retain, and learning new activities becomes more challenging.

As the disease progresses, disorientation becomes more prominent, leading to confusion about time, place, and even the people around them. Individuals may forget the day, month, or year, and this confusion can cause them to become lost, especially in unfamiliar environments. However, memories from the past—such as details about childhood, family, or significant historical events—tend to persist longer. While long-term memories may remain intact in the early stages, they worsen over time. Establishing routines can help individuals with Alzheimer's maintain some stability in their daily lives, offering a sense of structure as their memory declines. The loss of memory, combined with a decline in intellectual abilities, is crucial for diagnosing Alzheimer's, but other symptoms also emerge as the disease progresses.

Q9: What is the impact of Alzheimer's disease on language and communication?

Language impairment can be a significant symptom for individuals with Alzheimer's disease, sometimes appearing early alongside memory issues. Healthcare professionals assess various language challenges, including difficulties with word recall, comprehension, and writing.

Commonly, these language difficulties begin with word-finding problems (anomia), where individuals struggle to recall the correct word for objects, places, or events, or experience " tip of the tongue " moments. Naming objects correctly can also become challenging.

As the disease progresses, speech may become vague or less meaningful, even if sentences remain grammatically structured, particularly in the earlier stages. Individuals might make errors such as using newly invented words (neologisms, for example, calling a television a " watching box ") or substituting related but incorrect words (paraphasic errors, such as saying " fork " for " spoon "). The ability to understand spoken or written words can decline, and writing skills may also deteriorate.

These challenges understandably make communication more difficult. Forgetfulness and a potential loss of inhibitions can lead to conversational difficulties, such as repetitive behaviors or moments of uncharacteristic assertiveness. Interestingly, some research suggests that bilingual individuals may retain skills in a second language longer, which can sometimes facilitate better communication in their native or most frequently used language.

Studies indicate that while vocabulary and the ability to integrate meaning (semantics) are affected, the basic structure of sentences (syntax) may be relatively preserved, particularly in the earlier stages or in certain types of language tasks. For example, individuals might still produce grammatically correct sentences even if the overall meaning is unclear, or they may struggle to connect pronouns to references outside the immediate sentence.

Q10: How does Alzheimer's disease affect vision and spatial awareness?

As Alzheimer's disease progresses, individuals may struggle with recognizing objects, judging distances, or navigating their environment. These challenges stem from changes in the brain that affect areas responsible for visual processing and spatial orientation.

One common issue is visuospatial impairment, which can make it difficult to perceive and interpret visual information accurately. For example, a person with Alzheimer's might have trouble recognizing familiar faces or objects, even though their vision remains unchanged. They may confuse one object for another or fail to differentiate between objects that appear similar, leading to significant cognitive disorientation.

Spatial awareness is also impacted, making it harder for individuals to judge distances or navigate through spaces. This can result in difficulties with everyday tasks like driving, walking, or even moving around their home. For instance, a person may misjudge the distance to an object, leading to accidental bumps or trips over obstacles. In more severe cases, individuals with Alzheimer's may become disoriented and lost, even in familiar places, heightening the risks to their safety.

Q11: What is the difference between Alzheimer's disease and other types of dementia?

Dementias can be classified based on their underlying causes or the specific brain changes they produce. While each type of dementia has distinct pathological features and symptom patterns, there can often be overlapping characteristics. Alzheimer's disease, the most common form of dementia, is characterized by the accumulation of beta-amyloid plaques and tau protein tangles in the brain, starting with areas involved in memory, such as the hippocampus. As the disease progresses, these plaques and tangles interfere with neuronal communication, leading to the cognitive decline observed in Alzheimer's disease.

Other types of dementia are associated with different brain changes. Dementia with Lewy bodies (DLB), for example, involves the loss of brain cells and the formation of abnormal protein deposits known as Lewy bodies, which accumulate in regions like the limbic system and the neocortex. These deposits disrupt normal brain function, leading to symptoms such as visual hallucinations, movement problems, and cognitive fluctuations.

Frontotemporal dementia (FTD) primarily affects the frontal and temporal lobes of the brain, leading to specific molecular changes. FTD manifests in various forms, including progressive speech difficulties and significant personality or behavioral changes. This form of dementia is often diagnosed in younger individuals, typically before the age of 65.

Vascular dementia results from issues with blood flow in the brain, often caused by strokes or blocked blood vessels. The lack of adequate blood supply leads to cognitive decline, particularly in areas responsible for memory and decision-making.

While there are many other forms of dementia, Alzheimer's disease remains the most prevalent in older adults, making it the focus of the majority of scientific research and literature.

Q12: What are the differential diagnoses for Alzheimer's disease?

A differential diagnosis is the process of determining which condition is causing a set of symptoms, especially when different conditions might appear similar. Many health issues can cause symptoms resembling Alzheimer's disease, so it is important to rule out other possibilities. For example:

- *Vascular dementia,* which occurs when blood flow to the brain is reduced, often begins suddenly and progresses in a stepwise fashion.
- *Parkinson's disease dementia* involves memory problems but also includes symptoms like tremors and muscle stiffness.
- *Frontotemporal dementia* primarily affects personality and language more than memory and usually presents at a younger age than Alzheimer's.
- *Lewy body dementia* causes memory problems and may include hallucinations or motor symptoms similar to Parkinson's.
- *Depression* can lead to memory difficulties, but these often improve with treatment, whereas in Alzheimer's, memory problems worsen over time.
- *Normal pressure hydrocephalus (NPH),* which involves fluid buildup in the brain, can mimic Alzheimer's but can often be treated with surgery.
- *Thyroid disorders* or *vitamin deficiencies* can also cause memory loss, but treating these issues can improve symptoms.
- *Infections, brain injuries,* and conditions like *chronic traumatic encephalopathy* (CTE) can also cause similar cognitive problems.

A thorough medical exam, including tests and brain imaging, is necessary to ensure that Alzheimer's is correctly diagnosed and not another condition.

Q13: What is mild cognitive impairment (MCI)?

Mild cognitive impairment (MCI) is a condition that involves noticeable changes in mental abilities, such as memory, thinking, and reasoning skills. These changes are more pronounced than typical age-related changes but are not severe enough to interfere significantly with daily life and activities. People with MCI may experience memory lapses, such as frequently forgetting names or appointments, or may need help focusing, solving problems, or making decisions. Unlike dementia, MCI does not cause significant disruptions in daily living, and individuals with MCI are often still able to manage their care and function independently.

Different types of MCI are often categorized by the specific cognitive functions affected. For example, amnestic MCI primarily affects memory, while non-amnestic MCI impacts other mental skills, such as language or spatial abilities. MCI is considered a transitional stage between normal cognitive aging and dementia, particularly Alzheimer's disease. Not everyone with MCI will develop dementia; some people may remain stable, while others could even improve. However, individuals with MCI are generally at a higher risk of progressing to Alzheimer's or other forms of dementia, making early detection and monitoring essential for timely intervention and support.

Q14: How is mild cognitive impairment (MCI) related to Alzheimer's disease?

Mild cognitive impairment (MCI) is often seen as an intermediate stage between normal cognitive aging and Alzheimer's disease. While MCI does not always lead to dementia, individuals diagnosed with MCI, especially those with amnestic MCI (where memory issues are the main symptom), are at an elevated risk of progressing to Alzheimer's. Research suggests that people with MCI have a greater likelihood of developing Alzheimer's disease than those with no cognitive impairment, though the progression can vary widely among individuals.

In MCI, cognitive decline, especially memory problems, is more pronounced than what is typical with aging but does not significantly interfere with daily life or independence. In cases where MCI progresses to Alzheimer's, additional symptoms—such as increased memory loss, confusion, language difficulties, and impaired reasoning—usually develop over time, signaling the onset of more substantial cognitive decline.

Early detection of MCI allows for closer monitoring and potential interventions, which may delay progression to Alzheimer's and improve quality of life. This connection between MCI and Alzheimer's underscores the importance of recognizing early signs and differentiating between MCI and other age-related cognitive changes to manage health outcomes proactively.

Q15: What are the latest advances in Alzheimer's disease diagnosis as of 2024?

Significant advances are being made in Alzheimer's disease diagnosis, aiming for earlier and more accurate detection. Researchers are actively developing new tests, with a strong focus on biomarkers found in cerebrospinal fluid (CSF) and, increasingly, in blood. CSF biomarkers, such as specific forms of amyloid-beta and tau proteins, are more established and are used in specialist settings to aid in diagnosis, particularly in complex cases or for research.

Blood tests to detect these same proteins (e.g., plasma P-tau217) are a major area of development, offering the potential for less invasive and more accessible screening, possibly even before symptoms become apparent. While these blood-based biomarkers are rapidly advancing and show great promise for future clinical use, they are generally considered less validated for routine, standalone diagnosis compared to CSF tests or PET imaging and are often used currently as research tools or to help triage patients for further, more definitive testing. It's important to note that a positive biomarker test, whether from CSF, blood, or PET scan, indicates the presence of Alzheimer's pathology but must be interpreted within a full clinical context, as amyloid plaques can also be found in cognitively normal older adults.

Neuroimaging techniques like PET and MRI continue to evolve, providing more precise insights into brain changes. Amyloid PET and tau PET scans can visualize the hallmark plaques and tangles of Alzheimer's in the living brain. These are particularly useful for confirming amyloid pathology when considering new disease-modifying therapies.

While there is currently no cure for Alzheimer's disease, early and accurate diagnosis is crucial. It allows individuals and families to plan for the future, make informed decisions about medical care and finances, and access support services. Furthermore, early diagnosis is essential for eligibility for some of the newer disease-modifying treatments, such as anti-amyloid therapies, which aim to slow disease progression. However, it's important to maintain a balanced perspective: These treatments may offer modest benefits in slowing decline for some individuals and can have side effects; they do not reverse existing damage or stop the disease entirely. Early diagnosis also facilitates participation in clinical trials for emerging therapies.

Artificial intelligence and machine learning are also playing an increasing role, analyzing complex datasets from imaging, biomarkers, and clinical assessments to identify patterns indicative of the disease, further aiming to enhance diagnostic accuracy and facilitate timely intervention.

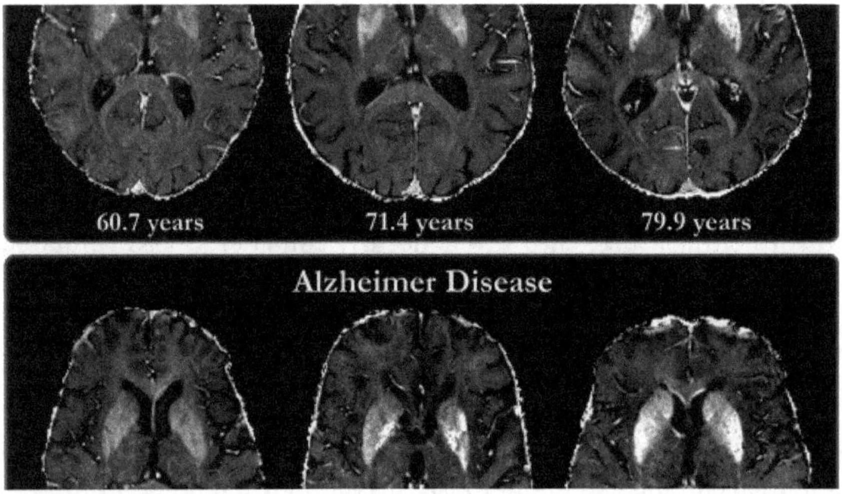

▲ FIGURE 3.5.
MRIs of aging patients with Alzheimer's disease over time.

Q16: How can family members recognize potential symptoms of Alzheimer's disease in their loved ones?

Family members may notice certain warning signs of Alzheimer's disease, including memory loss that disrupts daily life, such as forgetting recent conversations or events and asking the same questions repeatedly. Other signs include difficulty planning or solving problems, confusion with time or place, losing items and being unable to retrace steps, poor judgment, changes in mood or personality (such as becoming unusually anxious or withdrawn), and challenges in following conversations or completing familiar tasks. Subtle changes in behavior or interactions that family members recognize as "not typical" for their loved ones are often early indicators.

Q17: What physical tests are commonly used to assess symptoms of Alzheimer's disease?

Physical tests alone can't diagnose Alzheimer's, but they are part of a comprehensive diagnostic process to rule out other conditions. Typical assessments include blood tests to identify potential causes of symptoms (e.g., thyroid disorders, vitamin deficiencies), brain imaging (such as MRI, CT, or PET scans) to look for brain changes, like shrinkage in certain areas, and neurological exams that check reflexes, coordination, balance, and sensory response. These tests help identify any physical abnormalities and are often used alongside cognitive assessments.

Q18: What role do genetic tests play in diagnosing Alzheimer's disease?

Genetic tests play a complex role in Alzheimer's diagnosis. While they aren't used to definitively diagnose the disease, they can assess genetic risk factors, particularly the presence of the APOE-e4 allele, which is associated with a higher risk of developing Alzheimer's, especially late-onset. For early-onset Alzheimer's (occurring before age 65), genetic tests might check for mutations in genes such as APP, PSEN1, and PSEN2. However, genetic testing is not commonly recommended for everyone, as it can raise emotional and ethical considerations, and positive results do not guarantee the onset of the disease.

Q19: How do healthcare providers assess the rate of progression of Alzheimer's disease?

Healthcare providers assess the progression of Alzheimer's disease through regular cognitive assessments, monitoring behavioral and functional changes, and imaging techniques, which are sometimes used to observe structural changes in the brain over time. Tools like the Mini-Mental State Examination (MMSE) and the Montreal Cognitive Assessment (MoCA) can track cognitive decline. Additionally, standardized scales like the Clinical Dementia Rating (CDR) and the Functional Assessment Staging Test (FAST) help providers stage the disease and track changes in functional abilities, enabling them to understand the disease's impact on daily life and predict care needs.

Q20: What challenges exist in diagnosing Alzheimer's disease in younger patients?

Diagnosing Alzheimer's in younger patients (under 65) presents unique challenges, as symptoms are often attributed to stress, anxiety, or depression rather than dementia. These patients may also exhibit atypical symptoms, such as difficulties with language, decision-making, or spatial awareness, instead of the more classic memory-related symptoms. Since Alzheimer's is typically considered a condition of older adults, there is often a delay in considering it as a diagnosis. The rarity of early-onset Alzheimer's further complicates identification, increasing the likelihood of misdiagnosis or a prolonged diagnostic process.

Q21: How can early diagnosis impact the management of Alzheimer's disease?

Early diagnosis enables timely intervention, improving quality of life, aiding symptom management, and potentially slowing disease progression. It allows patients and families to plan for the future, make informed legal and financial decisions, and arrange for necessary support and care. Medications are often more effective when started early, and non-pharmacological interventions, such as cognitive therapies or lifestyle modifications (e.g., exercise and social engagement), can help preserve functioning for longer. Additionally, early diagnosis facilitates better access to clinical trials and research opportunities.

Q22: How often is Alzheimer's disease misdiagnosed, and what are common misdiagnoses?

Alzheimer's is misdiagnosed in 20–30% of cases, especially in early stages or younger patients. Common misdiagnoses include depression, anxiety, stress-related cognitive impairment, and other dementias such as frontotemporal or Lewy body dementia. Vascular dementia and mild cognitive impairment (MCI) are also often mistaken for Alzheimer's. Misdiagnosis arises because many Alzheimer's symptoms overlap with those of other neurological or psychiatric conditions, and a definitive diagnosis typically requires postmortem confirmation.

Q23: How does depression or anxiety sometimes mimic Alzheimer's symptoms?

Depression and anxiety can mimic cognitive impairment, presenting with symptoms like memory lapses, difficulty concentrating, and trouble making decisions. In severe cases, these conditions may also lead to confusion, lethargy, and disorientation, resembling early Alzheimer's—a phenomenon sometimes called *pseudo-dementia*. Depression, particularly in older adults, is also a recognized risk factor for cognitive decline. Addressing mental health symptoms is crucial for distinguishing cognitive disorders from treatable mental health issues. Comprehensive evaluations are essential to differentiate Alzheimer's from mood disorders accurately.

Q24: What is sundowning?

Sundowning refers to a set of symptoms often experienced by individuals with Alzheimer's disease and other forms of dementia, typically emerging in the late afternoon or evening. As the day progresses, those affected may exhibit increased confusion, agitation, anxiety, irritability, or restlessness. This behavior often contrasts with their daytime demeanor, which may be calmer and more oriented.

The exact cause of sundowning is not fully understood, but it is thought to be linked to factors such as fatigue, disruptions in the body's internal clock (circadian rhythms), and reduced natural light in the evening. Environmental factors like low lighting, routine changes, and increased shadows may also contribute to confusion and anxiety.

Managing sundowning often involves strategies such as:

- Establishing a consistent daily routine.
- Ensuring adequate lighting during the evening.
- Minimizing stressors.
- Engaging in calming activities to ease symptoms.

Q25: How does increasing neuroplasticity impact people with Alzheimer?

Increasing neuroplasticity—the brain's ability to reorganize and form new connections—offers promising potential for Alzheimer's patients. Enhanced neuroplasticity may allow the brain to compensate for the neuronal damage characteristic of Alzheimer's by creating alternative pathways to bypass affected areas. This adaptability can support cognitive functions such as memory, problem-solving, and decision-making, potentially slowing symptom progression, particularly in the disease's early stages. By strengthening neural networks, neuroplasticity could also provide a protective buffer, enabling individuals to maintain some degree of independence for longer.

In addition to cognitive benefits, greater neuroplasticity may foster emotional resilience in Alzheimer's patients. As the disease advances, patients often experience heightened anxiety, frustration, and depression. Enhanced neuroplasticity could allow the brain to adapt more effectively, helping patients cope with emotional challenges and reducing psychological distress. Moreover, it may amplify the effects of therapeutic interventions—such as cognitive exercises, physical activity, and social engagement—by increasing the brain's responsiveness to these activities. This heightened adaptability can improve patients' ability to retain new information and enhance their overall functional independence and emotional well-being.

Neuroplasticity also holds exciting possibilities for advancing Alzheimer's treatments. Research into interventions such as lifestyle modifications, cognitive rehabilitation, certain medications, and potentially psychedelics focuses on boosting neuroplasticity to manage symptoms rather than address the disease's underlying causes. By delaying the onset of severe symptoms, neuroplasticity-based strategies may alleviate caregiving demands and enhance patients' quality of life. While neuroplasticity cannot stop Alzheimer's progression, its capacity to help the brain adapt and maintain function underscores its importance as a promising focus in Alzheimer's care and research.

Q26: How does Alzheimer's disease affect daily living and activities?

Alzheimer's disease progressively impairs cognitive functions such as memory, reasoning, and executive skills, which are essential for daily activities. In the early stages, individuals may struggle with tasks like managing finances, following complex instructions, or organizing schedules. Middle stages often bring increased difficulty with routine activities such as cooking, cleaning, and personal hygiene. Late stages typically require full-time assistance for basic activities like bathing, dressing, and eating. Additionally, motor skills may decline, leading to challenges in walking or maintaining balance. The disease also affects the ability to recognize familiar places and people, increasing the risk of getting lost or confused in familiar environments. Caregivers often need to adapt to the living environment to enhance safety and support independence for as long as possible.

Q27: What are the early warning signs of Alzheimer's disease in younger adults?

In younger adults (typically under 65), early warning signs of Alzheimer's disease may include:

- *Subtle memory loss*: Difficulty remembering recent conversations, appointments, or where items are placed.
- *Language difficulties*: Struggling to find the right words during conversations or forgetting names of familiar objects.
- *Executive function decline*: Challenges with planning, organizing, multitasking, and problem-solving.
- *Visuospatial issues*: Trouble navigating familiar places or judging distances, which can affect driving.
- *Mood and personality changes*: Increased irritability, anxiety, depression, or unexplained mood swings.
- *Social withdrawal*: Losing interest in hobbies, social activities, or interacting with friends and family. These symptoms are often subtle and may be mistaken for stress, depression, or normal age-related changes, leading to delays in seeking medical evaluation.

Q28: Are there differences in symptoms between early-onset and late-onset Alzheimer's disease?

Yes, there are notable differences between early-onset (before age 65) and late-onset Alzheimer's disease:

- *Age of onset*: Early-onset typically affects individuals in their 40s to mid-60s, while late-onset occurs after 65.
- *Symptom progression*: Early-onset Alzheimer's often progresses more rapidly and may present with more pronounced non-memory symptoms, such as language impairments, visuospatial difficulties, and executive dysfunction.
- *Genetic factors*: Early-onset cases are more likely to have a genetic component, including mutations in genes like APP, PSEN1, and PSEN2.

- *Impact on career and family*: Early-onset can significantly disrupt careers and family life, as individuals are often still actively working and raising children.
- *Behavioral and psychological symptoms*: These may be more prominent in early-onset, including increased agitation, aggression, and mood disturbances. In contrast, late-onset Alzheimer's typically begins with memory impairment and progresses more slowly, with memory loss being the most prominent initial symptom.

Q29: How do mood changes present in individuals with Alzheimer's disease?

Mood changes are common in Alzheimer's disease and can manifest in various ways:

- *Depression*: Feelings of sadness, hopelessness, and a lack of interest in previously enjoyed activities.
- *Anxiety*: Excessive worry, fear, and nervousness, which may be triggered by confusion or inability to perform familiar tasks.
- *Irritability and agitation*: Increased frustration, impatience, and anger, often resulting from challenges in communication or understanding.
- *Apathy*: Reduced motivation, emotional flatness, and lack of initiative, making it difficult for individuals to engage in activities or interact with others.
- *Mood swings*: Rapid and unpredictable changes in mood, sometimes shifting from calmness to agitation without clear reason.
- *Social withdrawal*: Avoidance of social interactions and activities, leading to isolation and loneliness. These mood changes can be early indicators of Alzheimer's and may exacerbate cognitive symptoms by increasing stress and reducing the ability to cope with daily challenges.

Q30: What are the stages of Alzheimer's disease, and how do symptoms differ in each stage?

Alzheimer's disease is typically divided into three main stages: Early (Mild), Middle (Moderate), and Late (Severe). Each stage presents distinct symptoms:

- *Early Stage (Mild)*:
  - *Memory loss*: Difficulty remembering recent events, conversations, or appointments.

  - *Language problems*: Struggling to find the right words or repeating phrases.
  - *Impaired judgment*: Making poor decisions or experiencing difficulty with complex tasks.
  - *Mood changes*: Increased anxiety, depression, or irritability.
  - *Challenges with familiar tasks*: Trouble managing finances, driving, or organizing daily activities.
- *Middle Stage (Moderate)*:
  - *Increased memory loss*: Forgetting personal history, recognizing loved ones, or familiar places.
  - *Behavioral changes*: Increased agitation, aggression, wandering, or repetitive behaviors.
  - *Sleep disturbances*: Changes in sleep patterns, including insomnia or daytime drowsiness.
  - *Communication difficulties*: More pronounced language impairments and difficulty following conversations.
  - *Assistance with daily activities*: Need for help with dressing, bathing, and eating.
- *Late Stage (Severe)*:
  - *Profound memory loss*: Inability to recognize family members or recall personal history.
  - *Loss of motor skills*: Difficulty walking, sitting, and eventually swallowing.
  - *Communication loss*: Minimal verbal communication; reliance on non verbal cues.
  - *Total dependency*: Complete reliance on caregivers for all daily needs.
  - *Vulnerability to infections*: Increased susceptibility to pneumonia and other infections due to weakened immune response and immobility.

Understanding these stages helps caregivers and healthcare providers anticipate changes, plan appropriate care, and implement strategies to improve the quality of life for individuals with Alzheimer's disease.

Q31: How do symptoms vary between individuals with Alzheimer's disease?

Symptoms of Alzheimer's disease can vary widely among individuals due to factors such as age, genetics, overall health, and the presence of other medical conditions. Key variations include:

- *Onset and progression*: Some individuals experience a rapid decline, while others progress more slowly.
- *Primary symptoms*: While memory loss is common, some may exhibit more pronounced language difficulties, visuospatial impairments, or behavioral changes.
- *Cognitive domains affected*: Variability in which cognitive functions are most impacted, such as executive function, attention, or motor skills.
- *Psychological symptoms*: Differences in mood and behavior, with some individuals experiencing significant depression or anxiety, while others may show increased agitation or apathy.
- *Response to treatment*: Varied responses to medications and therapeutic interventions, influencing symptom management and quality of life.
- *Coexisting conditions*: Presence of other health issues, such as cardiovascular disease or diabetes, can influence the presentation and progression of Alzheimer's symptoms. These variations necessitate personalized care plans and tailored interventions to address the unique needs of each individual.

Q32: What role do neuroimaging techniques (e.g., MRI, PET scans) play in diagnosing Alzheimer's disease?

Neuroimaging techniques are crucial in diagnosing Alzheimer's disease by providing detailed images of the brain's structure and function:

- *Magnetic Resonance Imaging (MRI)*:
  - Structural MRI assesses brain atrophy, particularly in the hippocampus and cortical regions, which are indicative of Alzheimer's.
  - Differential diagnosis: Helps exclude other causes of dementia, such as tumors, strokes, or vascular changes.
- *Positron Emission Tomography (PET) Scans*:
  - Amyloid PET detects amyloid-beta plaques, a hallmark of Alzheimer's, by using radioactive tracers that bind to these proteins.
  - Tau PET identifies tau tangles, another key feature of Alzheimer's, providing insights into disease progression.
  - FDG-PET measures glucose metabolism in the brain, with reduced uptake in specific regions suggesting Alzheimer's.

- *Functional MRI (fMRI)*: Assesses brain activity by measuring blood flow, helping to identify functional impairments related to cognitive tasks.
- *Single-Photon Emission Computed Tomography (SPECT)*: Evaluates cerebral blood flow patterns, which can show reductions in areas typically affected by Alzheimer's. Neuroimaging not only aids in confirming a diagnosis but also assists in staging the disease, monitoring progression, and evaluating the effectiveness of treatments.

Q33: How do biomarkers in cerebrospinal fluid (CSF) or blood aid in Alzheimer's diagnosis?

Biomarkers in CSF and blood provide objective measures to support the diagnosis of Alzheimer's disease:

- *Cerebrospinal Fluid (CSF) biomarkers*:
  - Amyloid-beta (Aβ42): Reduced levels in CSF indicate amyloid plaque deposition in the brain.
  - Total Tau (t-tau) and Phosphorylated Tau (p-tau): Elevated levels reflect neuronal damage and tau pathology.
- *Blood biomarkers*:
  - Plasma amyloid-beta: Emerging blood tests can detect changes in amyloid-beta levels similar to CSF measurements.
  - Plasma Tau: Recent advancements allow for the detection of tau proteins in blood, correlating with neurodegeneration.
  - Neurofilament Light Chain (NfL): Elevated in both CSF and blood, indicating neuronal injury and disease progression.
- *Genetic biomarkers*:
  - Apolipoprotein E (APOE) ε4 allele: Genetic testing for APOE ε4 can assess the risk and potential progression of Alzheimer's, though it is not definitive for diagnosis.
- *Advanced techniques*:
  - Exosomes: Tiny vesicles in blood that carry brain-derived proteins, offering potential for non invasive biomarker discovery. Biomarkers enhance diagnostic accuracy, allow for earlier detection (even before significant cognitive symptoms emerge), and

facilitate the differentiation of Alzheimer's from other neurodegenerative diseases. They also play a vital role in research and the development of targeted therapies.

Q34: What psychological assessments are used to diagnose Alzheimer's disease?

Psychological assessments are integral to diagnosing Alzheimer's disease by evaluating cognitive and functional impairments:

- *Mini-Mental State Examination (MMSE)*:
  - Scope: Assesses orientation, registration, attention, calculation, recall, language, and visuospatial skills.
  - Usage: Commonly used as a screening tool to identify cognitive impairment and monitor changes over time.
- *Montreal Cognitive Assessment (MoCA)*:
  - Scope: Evaluates executive functions, attention, memory, language, visuospatial skills, and orientation.
  - Advantages: More sensitive than MMSE for detecting mild cognitive impairment.
- *Clock drawing test*:
  - Purpose: Assesses visuospatial and executive functioning by having the patient draw a clock showing a specific time.
  - Interpretation: Errors may indicate cognitive deficits related to Alzheimer's.
- *Neuropsychological testing*:
  - Comprehensive evaluation: Includes a battery of tests to assess various cognitive domains in detail, such as memory, attention, language, executive function, and visuospatial abilities.
  - Customization: Tailored to the individual's specific needs and symptoms for a thorough assessment.
- *ADAS-Cog (Alzheimer's Disease Assessment Scale-Cognitive Subscale)*:
  - Scope: Measures memory, language, praxis, and orientation.
  - Usage: Often used in clinical trials to evaluate the efficacy of treatments.
- *Functional assessments*:
  - Activities of Daily Living (ADL): Evaluates the ability to perform basic tasks like bathing, dressing, and eating.

- Instrumental Activities of Daily Living (IADL): Assesses more complex activities such as managing finances, medication, and household chores. These assessments help clinicians determine the presence and severity of cognitive impairments, differentiate Alzheimer's from other forms of dementia, and develop appropriate care plans.

Q35: How do physicians distinguish Alzheimer's disease from normal aging?

Physicians differentiate Alzheimer's disease from normal aging by evaluating the extent, type, and impact of cognitive changes:

- *Memory impairment*:
  - Normal aging: Occasional forgetfulness, such as misplacing keys or forgetting names, without significant impact on daily life.
  - Alzheimer's: Persistent and progressive memory loss that interferes with daily activities, such as forgetting important dates, repeatedly asking the same questions, or relying heavily on memory aids.
- *Cognitive decline*:
  - Normal aging: Slower processing speed and minor difficulties with multitasking, which do not significantly affect functionality.
  - Alzheimer's: Noticeable decline in cognitive functions, including impaired judgment, problem-solving, language difficulties, and reduced ability to perform complex tasks.
- *Functional impact*:
  - Normal aging: Cognitive changes do not substantially hinder independence or the ability to manage personal and professional responsibilities.
  - Alzheimer's: Significant impairment in daily functioning, requiring assistance with routine activities and increasing dependency on others.
- *Behavioral and psychological symptoms*:
  - Normal aging: Generally stable mood and behavior with occasional stress-related changes.
  - Alzheimer's: Frequent mood swings, depression, anxiety, agitation, and changes in personality.
- *Diagnostic tools*:
  - Cognitive assessments: Scores on tests like MMSE or MoCA are significantly lower in Alzheimer's compared to normal aging.

- Neuroimaging and biomarkers: Presence of amyloid plaques, tau tangles, and brain atrophy distinguish Alzheimer's from normal aging. By combining clinical evaluations, cognitive testing, and diagnostic imaging, physicians can accurately distinguish Alzheimer's disease from the cognitive changes associated with normal aging.

Q36: How reliable are current diagnostic methods for Alzheimer's disease?

Current diagnostic methods for Alzheimer's disease have significantly improved in reliability due to advancements in technology and a better understanding of the disease. Key factors influencing reliability include:

- *Clinical assessments*:
  - Cognitive tests: Tools like MMSE and MoCA have high sensitivity and specificity, especially when combined with other diagnostic methods.
  - Functional assessments: Evaluations of daily living activities enhance diagnostic accuracy by correlating cognitive deficits with functional impairments.
- *Neuroimaging*:
  - MRI and CT scans: Effectively rule out other causes of dementia and identify characteristic brain atrophy patterns, increasing diagnostic confidence.
  - PET scans: High specificity in detecting amyloid and tau pathologies, making them reliable for confirming Alzheimer's in conjunction with clinical symptoms.
- *Biomarkers*:
  - CSF and blood tests: Biomarkers like Aβ42, t-tau, and p-tau have shown high accuracy in identifying Alzheimer's, especially when used alongside imaging and clinical assessments.
  - Genetic testing: Identifies risk factors (e.g., APOE ε4) but is less definitive for diagnosis.
- *Standardized criteria*:
  - NIA-AA and IWG Guidelines: Provide comprehensive frameworks that integrate clinical, imaging, and biomarker data, enhancing diagnostic consistency and reliability.
- *Multidisciplinary approach*:
  - Collaborative evaluations: Involving neurologists, psychiatrists, and neuropsychologists ensures a

thorough and accurate diagnosis by considering multiple perspectives. Despite these advancements, some challenges remain:

- *Overlap with other dementias*: Similar symptoms can make differential diagnosis difficult without comprehensive testing.
- *Accessibility of advanced diagnostics*: PET scans and biomarker tests may not be readily available in all healthcare settings.
- *Early detection limitations*: While biomarkers and imaging are improving, early-stage Alzheimer's can still be challenging to diagnose definitively before significant symptoms appear. Overall, while no single diagnostic method is perfect, the combination of clinical assessments, neuroimaging, and biomarker testing provides a highly reliable framework for diagnosing Alzheimer's disease.

–

In conclusion, early recognition of the symptoms and timely diagnosis of Alzheimer's disease are critical for managing the condition effectively. By distinguishing between normal aging and the early signs of Alzheimer's, individuals and healthcare professionals can take proactive steps to address cognitive, behavioral, and physical changes. The diagnostic process, encompassing medical evaluations, cognitive assessments, and advanced imaging, provides a comprehensive approach to confirming the presence of Alzheimer's and understanding its progression. Early intervention not only helps in managing symptoms but also opens the door to treatment options and support resources that can enhance quality of life. With a thorough understanding of the indicators and diagnostic methods, patients and caregivers are better equipped to navigate the journey of Alzheimer's with greater awareness and preparation.

Bibliography

1. ***Bouchie, G. (n.d.).*** *Brain fingerprints: Detecting alzheimer's disease early with Stephanie Aghamoosa, Phd. MUSC Research. https://research.musc.edu/stories/science-never-sleeps/2024/april/detecting-alzheimers*
2. ***Budson, A. E., & Solomon, P. R.*** *(2016). Memory loss, Alzheimer's disease, and dementia: a practical guide for clinicians (Second edition). Elsevier.*
3. ***Chiaravalloti, N. D., & Goverover, Yael.*** *(Eds.). (2017). Changes in the Brain : Impact on Daily Life (1st ed.). Springer New York. https://doi.org/10.1007/978-0-387-98188-8*

4. **Cusimano, C.** *(2021). Language and neurology: Alzheimer's disease. ISTE Ltd.*
5. *Diagnosis of dementia due to Alzheimer's disease: Recommendations from the National Institute on Aging-Alzheimer's Association workgroups on diagnostic guidelines for Alzheimer's disease. Alzheimers Dement. 2011;7(3):263-9.*
6. **Dubois B, Feldman HH, Jacova C, Hampel H, Molinuevo JL, Blennow K, et al.** *Advancing research diagnostic criteria for Alzheimer's disease: The IWG-2 criteria. Lancet Neurol. 2014;13(6):614-29.*
7. **Esther Heerema, M.** *(2025, January 16). Communication and language difficulties in alzheimer's disease. Verywell Health. https://www.verywellhealth.com/alzheimers-disease-and-word-finding-difficulties-98554#:~:text=Word%2Dfinding%20aphasia%20may%20manifest,understanding%20of%20spoken%20or%20written*
8. **Fiveable.** *(n.d.). Syntax - (intro to cultural anthropology) - vocab, definition, explanations: Fiveable. All 38 AP Subjects. https://library.fiveable.me/key-terms/introduction-cultural-anthropology/syntax*
9. **Gruetzner, H.** *(2001). Alzheimer's: a caregiver's guide and sourcebook (3rd ed.). Wiley.*
10. **Jessen F, Amariglio RE, van Boxtel M, Breteler M, Ceccaldi M, Chételat G, et al.** *A conceptual framework for research on subjective cognitive decline in preclinical Alzheimer's disease. Alzheimers Dement. 2014;10(6):844-52.*
11. **University of California San Francisco. (n.d.).** *Speech & Language. Memory and Aging Center. https://memory.ucsf.edu/brain-health/speech-language#:~:text=Fluent%2C%20grammatically%20correct%20speech%20with,a%20%E2%80%9Cfork%E2%80%9D%20(semantic)*

CHAPTER 4

Treatments and Therapies

While Alzheimer's disease currently has no cure, significant advancements in medical research and therapeutic interventions provide a foundation for improving patients' quality of life, slowing disease progression, and addressing associated symptoms. This chapter begins by examining FDA-approved medications that target memory loss, cognitive decline, and behavioral symptoms, such as cholinesterase inhibitors and NMDA receptor antagonists. It also explores emerging treatments, including monoclonal antibody therapies and other innovations aimed at targeting the underlying biology of Alzheimer's, such as amyloid plaques and tau proteins.

Beyond pharmacological options, the chapter highlights the importance of non drug interventions, including cognitive therapy, physical exercise, and social engagement, which have been shown to enhance mental and emotional well-being. Lifestyle modifications, such as adopting a brain-healthy diet and managing cardiovascular risk factors, are discussed as essential components of preventive care. Additionally, this chapter underscores the critical role of caregiver education and support systems in creating a nurturing environment for those affected by the disease. By presenting a comprehensive overview of these treatments and therapies, this chapter equips readers with the knowledge to navigate the available options and make informed decisions tailored to the needs of individuals living with Alzheimer's.

–

Q1: What are the current treatments for Alzheimer's disease?

Current treatments for Alzheimer's disease primarily focus on managing symptoms and slowing disease progression. Approved medications include cholinesterase inhibitors (donepezil, rivastigmine, and galantamine) and the NMDA receptor antagonist memantine. These drugs can improve cognitive function and reduce behavioral symptoms but do not provide a cure.

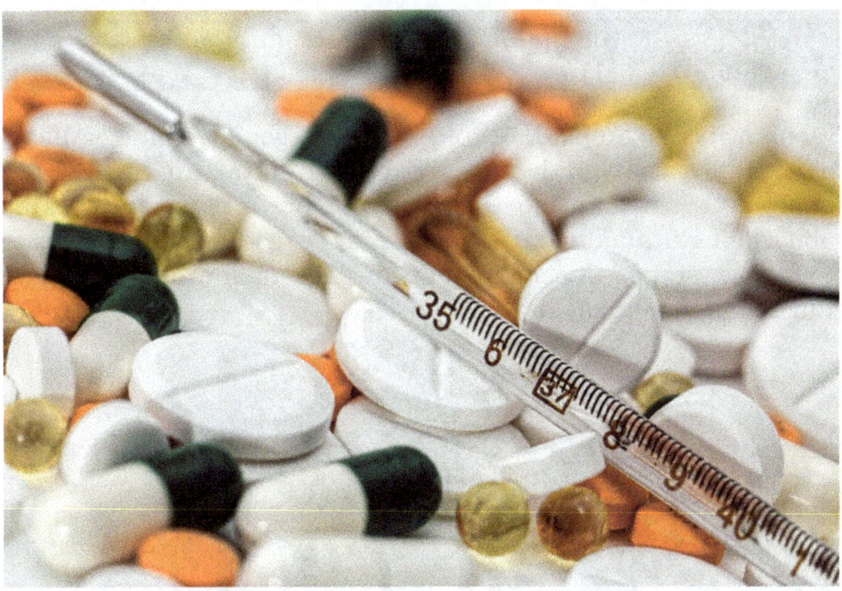

▲ **FIGURE 4.1.**
Thermometer on medical pills.

Q2: How do cholinesterase inhibitors work in Alzheimer's disease?

Cholinesterase inhibitors, such as donepezil, rivastigmine, and galantamine, prevent the breakdown of acetylcholine, a neurotransmitter essential for learning and memory. By increasing acetylcholine levels in the brain, these medications improve cognitive function and may slow the progression of Alzheimer's symptoms.

▲ **FIGURE 4.2.**
An elderly man exercising.

Q3: What is the role of memantine in treating Alzheimer's disease?

Memantine is an NMDA receptor antagonist that regulates glutamate activity in the brain. Excessive glutamate can damage neurons, so memantine helps protect brain cells by blocking abnormal glutamate activity. It is often prescribed for moderate to severe stages of Alzheimer's disease to improve cognitive function and support daily activities.

▲ **FIGURE 4.3.**
An elderly man enjoys regular activities with medicinal support.

Q4: What are the potential side effects of Alzheimer's disease medications?

Potential side effects of Alzheimer's disease medications vary depending on the drug. Cholinesterase inhibitors may cause nausea, vomiting, diarrhea, and loss of appetite. In contrast, memantine may lead to dizziness, headache, constipation, and confusion. Healthcare providers must closely monitor patients and adjust treatments to minimize side effects.

Q5: Are there any disease-modifying treatments for Alzheimer's disease?

There are no fully effective disease-modifying treatments for Alzheimer's disease, but research is ongoing. Some experimental drugs aim to target amyloid plaques and tau tangles to slow disease progression. Aducanumab, a monoclonal antibody targeting amyloid, has shown promise and received FDA approval for early Alzheimer's; however, its effectiveness and usage are still under evaluation.

▲ **FIGURE 4.4.**
Memantine.

Q6: How do non-pharmacological therapies help in managing Alzheimer's disease?

Non-pharmacological therapies, including cognitive stimulation, physical exercise, social engagement, and lifestyle modifications, play a crucial role in managing Alzheimer's disease. These approaches can help maintain cognitive function, improve mood, and enhance the quality of life for individuals with the condition. Common activities include puzzles, art therapy, music therapy, and social interaction.

Q7: What is the role of cognitive stimulation in Alzheimer's disease management?

Cognitive stimulation involves activities that challenge the brain and promote mental agility, such as puzzles, games, reading, and learning new skills. Engaging in these activities can help maintain cognitive function and may potentially slow the progression of Alzheimer's disease. Programs and therapies focused on mental training and rehabilitation serve as valuable tools for supporting individuals with the condition.

Q8: How does physical exercise benefit individuals with Alzheimer's disease?

Physical exercise benefits individuals with Alzheimer's disease by improving cardiovascular health, enhancing mood, and supporting cognitive function. Regular exercise can help reduce symptoms of depression and anxiety, improve the quality of sleep, and maintain physical mobility. Walking, swimming, and strength training can be adapted to suit different fitness levels and abilities.

Q9: What dietary approaches are beneficial for Alzheimer's disease?

Dietary approaches beneficial for Alzheimer's include the Mediterranean and MIND diets, which emphasize fruits, vegetables, whole grains, lean proteins, and healthy fats. These diets are rich in antioxidants, omega-3 fatty acids, and other nutrients that support brain health. Avoiding processed foods, excessive sugar, and unhealthy fats is crucial for maintaining overall health and cognitive function.

Q10: What is the impact of music therapy on Alzheimer's disease?

Music therapy positively impacts individuals with Alzheimer's disease by improving mood, reducing agitation, and enhancing cognitive function. Listening to music, singing, and playing instruments can stimulate the brain, evoke memories, and provide comfort and joy. Music therapy often involves engaging individuals with Alzheimer's in meaningful and enjoyable activities.

Q11: How do behavioral interventions help manage Alzheimer's disease symptoms?

Behavioral interventions help manage Alzheimer's disease symptoms by addressing specific challenges such as agitation, aggression, and sleep disturbances. Strategies include creating a calm environment, using distraction and redirection, establishing routines, and providing reassurance. Tailoring interventions to individual preferences and needs is essential for their effectiveness.

Q12: What is the role of occupational therapy in Alzheimer's disease care?

Occupational therapy plays a crucial role in Alzheimer's disease care by helping individuals maintain independence and perform daily activities. Occupational therapists assess a person's abilities and provide strategies and tools to support self-care, mobility, and safety. They may also recommend environmental modifications to enhance functioning and improve quality of life.

Q13: How does speech therapy assist individuals with Alzheimer's disease?

Speech therapy helps individuals with Alzheimer's by addressing communication and swallowing challenges. Speech therapists focus on improving language skills, such as word-finding and comprehension, and provide exercises to enhance speech clarity. They also develop strategies to manage dysphagia (difficulty swallowing) and promote safe, adequate nutrition.

Q14: What is the role of support groups for Alzheimer's disease patients and caregivers?

Support groups are essential in offering emotional and practical support to individuals with Alzheimer's disease and their caregivers. These groups create a safe space for sharing experiences, obtaining valuable information, and finding encouragement. By reducing feelings of isolation, providing effective coping strategies, and fostering a sense of community, support groups help participants navigate the challenges of Alzheimer's disease more effectively.

Q15: How do pet therapy and animal-assisted interventions benefit individuals with Alzheimer's?

Pet therapy and animal-assisted interventions offer significant benefits for individuals with Alzheimer's disease by alleviating stress, boosting mood, and promoting social interaction. The companionship of animals, such as therapy dogs, provides comfort, emotional support, and a sense of purpose. Engaging with animals can also stimulate memory, encourage communication, and foster meaningful connections for those living with Alzheimer's.

Q16: What are the emerging therapies and treatments for Alzheimer's disease?

Emerging therapies for Alzheimer's disease include innovative drugs targeting amyloid plaques and tau tangles, as well as advancements in gene therapy and immunotherapy. Progress in biomarkers and diagnostic tools is enabling earlier detection and more timely interventions. Additionally, non-pharmacological approaches, such as personalized cognitive training and lifestyle modifications, are being actively researched for their potential to enhance patient outcomes.

Q17: How does gene therapy work for Alzheimer's disease?

Gene therapy for Alzheimer's focuses on delivering genetic material to the brain to modify specific genes linked to the disease. This innovative approach aims to decrease the production of harmful proteins, boost the brain's repair mechanisms, or introduce protective factors. Currently, gene therapy remains in the experimental phase, with ongoing research dedicated to evaluating its safety and effectiveness.

Q18: What is the potential of immunotherapy in treating Alzheimer's disease?

Immunotherapy offers a promising approach to treating Alzheimer's disease by leveraging the body's immune system to target and eliminate amyloid plaques and tau tangles. Strategies such as monoclonal antibodies, vaccines, and other immunotherapeutic methods are under development and undergoing clinical trials. While initial results are encouraging, further research is essential to establish their long-term effectiveness and safety.

Q19: How does transcranial magnetic stimulation (TMS) help in Alzheimer's disease?

Transcranial magnetic stimulation (TMS) helps with Alzheimer's disease by using magnetic fields to stimulate nerve cells in the brain. TMS can enhance cognitive function, improve mood, and reduce symptoms of depression in individuals with Alzheimer's. It is a non invasive procedure that is being explored as a potential treatment to support brain health and cognitive performance.

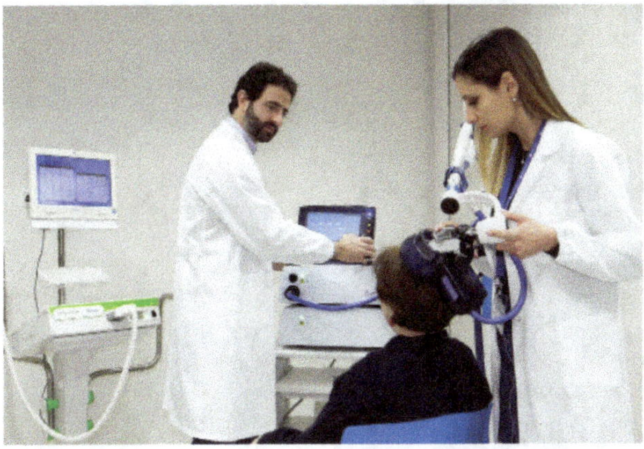

▲ **FIGURE 4.5.**
TMS scan.

Q20: What are the challenges in developing new treatments for Alzheimer's disease?

Developing new treatments for Alzheimer's disease faces several challenges, including the disease's complex and multifactorial nature, the difficulty of crossing the blood–brain barrier, and variability in patient responses. High costs, lengthy timelines for clinical trials, and the need for early and accurate diagnosis also pose significant obstacles. Collaborative efforts, innovative research approaches, and increased funding are essential to overcoming these challenges and advancing the development of effective treatments.

Q21: What is the role of stem cell therapy in Alzheimer's disease treatment?

Stem cell therapy holds promise for Alzheimer's disease treatment by offering the potential to replace damaged neurons, regenerate brain tissue, and modulate the immune response. Stem cells can be derived from various sources, including embryonic tissue, adult tissues like bone marrow and adipose tissue, and induced pluripotent stem cells (iPSCs). Research is ongoing to explore the safety and efficacy of various stem cell-based approaches in preclinical and clinical studies.

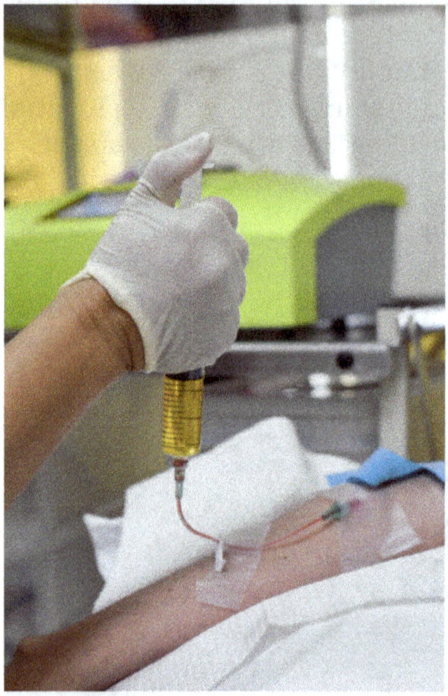

▲ **FIGURE 4.6.**
Stem cell therapy.

Q22: How does epigenetics influence Alzheimer's disease risk and progression?

Epigenetics, which refers to changes in gene expression that do not involve alterations in the DNA sequence itself, plays a significant role in Alzheimer's disease risk and progression. Environmental factors such as diet, stress, and exposure to toxins can modify gene expression patterns, affecting processes like inflammation, neuronal function, and amyloid metabolism. Understanding epigenetic mechanisms in Alzheimer's may lead to new therapeutic targets and personalized treatment approaches.

Q23: What are the challenges in conducting clinical trials for Alzheimer's disease treatments?

Clinical trials for Alzheimer's disease treatments face several challenges, including recruiting and retaining participants, ensuring diverse representation, and selecting appropriate outcome measures. The disease's heterogeneity, with multiple underlying pathologies and variable symptom profiles, complicates trial design and the interpretation of results. Longitudinal studies with large sample sizes and innovative trial designs are needed to overcome these challenges and accelerate the development of effective treatments.

Q24: How does precision medicine approach Alzheimer's disease treatment?

Precision medicine aims to tailor healthcare interventions to individual characteristics, including genetic makeup, biomarker profiles, and lifestyle factors. In Alzheimer's, precision medicine approaches seek to identify disease subtypes with distinct underlying mechanisms and treatment responses. Personalized treatment plans may involve targeted therapies based on genetic risk factors, biomarker profiles, and other customized factors to optimize outcomes for individuals with Alzheimer's.

Q25: What are the ethical considerations in Alzheimer's disease research and treatment?

Alzheimer's disease research and treatment raise ethical considerations regarding informed consent, privacy, autonomy, and equity. Issues such as genetic testing, the use of biomarkers for diagnosis, and end-of-life care require careful consideration of individual rights, safety, and societal values. Ethical frameworks and guidelines help ensure that research and clinical practices uphold moral principles and respect the dignity and well-being of individuals with Alzheimer's and their caregivers.

Q26: How does artificial intelligence (AI) contribute to Alzheimer's disease research and treatment?

Artificial intelligence (AI) contributes to Alzheimer's disease research and treatment by analyzing large datasets, identifying patterns, and predicting disease progression. AI algorithms can analyze neuroimaging data, genetic profiles, and clinical records to improve early detection, diagnostic accuracy, and treatment selection. Machine learning techniques also help uncover novel biomarkers, drug targets, and therapeutic strategies, accelerating progress toward effective treatments for Alzheimer's.

Q27: What are the societal impacts of Alzheimer's disease?

Alzheimer's disease has significant societal impacts, including healthcare costs, caregiver burden, and loss of productivity. As the population ages, the prevalence of Alzheimer's is expected to increase, placing a growing strain on healthcare systems and social support networks. Addressing the societal impacts of Alzheimer's requires investment in research, education, caregiving resources, and policies that promote dementia-friendly communities and support systems.

Q28: How does Alzheimer's disease affect caregivers?

Alzheimer's disease affects caregivers physically, emotionally, and financially. Caregivers often experience high levels of stress, anxiety, depression, and social isolation as they provide care for their loved ones with Alzheimer's. Balancing caregiving responsibilities with work and personal life can be challenging, leading to caregiver burnout and health problems. Support services, respite care, and caregiver education programs are essential for alleviating caregiver burden and promoting well-being.

Q29: What are the economic costs of Alzheimer's disease?

Alzheimer's disease imposes a profound economic burden on individuals, families, healthcare systems, and society. Direct costs include medical expenses for diagnosis, treatment, and long-term care, while indirect costs stem from lost productivity, caregiver burden, and reduced quality of life. In the United States alone, the annual cost of Alzheimer's care exceeds hundreds of billions of dollars. As the global population ages and the prevalence of the disease rises, these costs are projected to escalate significantly, highlighting the urgent need for effective prevention and treatment strategies to mitigate its impact on healthcare systems and economies worldwide.

Q30: How can communities become more dementia-friendly?

Communities can become more dementia-friendly by raising awareness, reducing stigma, and providing support services for individuals with Alzheimer's and their caregivers. Dementia-friendly initiatives involve:

- Educating the public, businesses, and organizations about dementia.

- Adapting physical environments to be more accessible and inclusive.
- Offering dementia-friendly programs and activities.

Collaboration between stakeholders, including government agencies, healthcare providers, and community groups, is essential for creating dementia-friendly communities that support the well-being and independence of individuals with Alzheimer's.

Q31: How effective are current Alzheimer's treatments in slowing disease progression?

Current treatments, including cholinesterase inhibitors and memantine, help manage symptoms but generally have limited effects on slowing the disease's progression. They may improve cognitive and functional abilities for a period, especially in mild to moderate stages, but they do not stop or reverse the disease. Clinical studies continue to explore disease-modifying treatments that might slow the underlying progression of Alzheimer's.

Q32: How do combination therapies work in treating Alzheimer's disease?

Combination therapies involve using multiple drugs or a mix of pharmacological and non-pharmacological approaches to address Alzheimer's from various angles. For instance, cholinesterase inhibitors combined with memantine may be used in moderate to severe stages. Non drug therapies, such as cognitive training combined with medication, may also enhance quality of life and functional abilities. However, effectiveness varies by individual, and more research is needed on the benefits of specific combinations.

Q33: What role does light therapy play in Alzheimer's disease treatment?

Light therapy, or bright light exposure, improves sleep patterns and reduces behavioral symptoms such as agitation and confusion, often seen in Alzheimer's patients. Exposure to bright, natural light in the morning can help regulate circadian rhythms, potentially improving sleep quality and mood. While not a cure, it can be a helpful addition to the treatment plan, especially for those experiencing severe sleep disturbances.

Q34: How can acupuncture and traditional Chinese medicine contribute to Alzheimer's disease management?

Acupuncture and specific herbal remedies in traditional Chinese medicine are believed to improve blood flow and reduce inflammation, which may positively affect cognitive function. While studies are limited, some show mild improvements in memory and anxiety in Alzheimer's patients receiving acupuncture. Caution is advised, as herbal supplements can interact with other medications, so healthcare professionals should oversee these treatments.

Q35: What complementary therapies, such as mindfulness or yoga, benefit Alzheimer's patients?

Complementary therapies like mindfulness meditation, yoga, and breathing exercises can help reduce stress, anxiety, and depression in Alzheimer's patients. These practices may improve overall well-being, mood, and sleep quality, offering a gentle and non invasive approach to symptom management. They are also beneficial for caregivers, helping them manage stress and maintain a positive outlook.

Q36: How does neuroplasticity-based therapy support cognitive function in Alzheimer's?

Neuroplasticity-based therapies, like cognitive training exercises, strengthen existing neural connections and stimulate brain activity. By challenging cognitive skills, these therapies may help slow cognitive decline in the early stages of Alzheimer's. Techniques such as memory games, puzzles, and other structured mental activities encourage the active engagement of cognitive pathways, helping to maintain functionality for extended periods.

Q37: What are the risks and benefits of off-label medications in Alzheimer's treatment?

Off-label medications, like certain antidepressants or antipsychotics, can help manage behavioral symptoms in Alzheimer's patients, such as aggression or severe anxiety. However, these drugs carry risks, especially in elderly patients, including increased chances of falls, cardiovascular issues, and sometimes worsened cognitive symptoms. They should be used cautiously and only when behavioral symptoms significantly impact quality of life.

Q38: How does art therapy improve the quality of life in Alzheimer's patients?

Art therapy provides a creative outlet, which can help reduce anxiety, improve mood, and enhance self-expression in Alzheimer's patients, even when verbal communication is limited. Engaging in artistic activities can stimulate brain function, offering moments of clarity and enjoyment. Art therapy also promotes social interaction and enhances emotional connection with caregivers.

Q39: How is pharmacogenetics used to personalize Alzheimer's disease treatments?

Pharmacogenetics tailors drug treatments based on a patient's genetic profile, helping predict how they may respond to specific medications. In Alzheimer's, this approach could improve drug effectiveness and minimize adverse effects by identifying the best medications and dosages for each patient. While still in the early stages of research, pharmacogenetics may become increasingly relevant as more Alzheimer's drugs are developed.

Q40: What is the impact of neurofeedback in managing Alzheimer's symptoms?

Neurofeedback, a form of biofeedback that uses real-time monitoring of brain activity, aims to improve cognitive function by training the brain to regulate its activity. Though research is limited, some studies suggest it may help improve focus, memory, and mood in Alzheimer's patients. More robust evidence is needed to confirm its effectiveness as a regular therapy for Alzheimer's disease.

Q41: How do wearable devices support Alzheimer's disease management?

Wearable devices, like smartwatches or fitness trackers, can monitor activity levels, heart rate, sleep patterns, and even track location, providing valuable data to healthcare providers and caregivers. These devices can help ensure safety, detect routine changes, and remind patients to take medications, thus enhancing overall management and quality of life.

Q42: How can nutritional supplements like omega-3 fatty acids help with Alzheimer's symptoms?

Omega-3 fatty acids, found in fish oil, are thought to support brain health and may slow cognitive decline in some Alzheimer's patients, although the evidence is mixed. Other supplements, such as antioxidants and B vitamins, are also being studied for potential cognitive benefits. However, patients should consult healthcare providers before taking supplements to avoid interactions with medications.

Q43: What are the effects of aromatherapy on mood and anxiety in Alzheimer's patients?

Aromatherapy, often using essential oils like lavender or rosemary, can have calming effects, reducing anxiety and agitation in Alzheimer's patients. Diffusing oils or applying them in diluted forms during massage can help soothe patients and create a relaxing environment, though effects may vary. Aromatherapy is a safe and non invasive complementary therapy but should be used cautiously to avoid allergies.

Q44: What role do anti-inflammatory drugs have in treating or preventing Alzheimer's?

Chronic inflammation is implicated in Alzheimer's progression, and anti-inflammatory drugs have been explored as potential treatments. Non-steroidal anti-inflammatory drugs (NSAIDs), like ibuprofen, showed initial promise but have not proven effective in clinical trials for Alzheimer's prevention or treatment. Research continues to explore new anti-inflammatory drugs with a more targeted approach to brain inflammation.

Q45: How are antioxidants and neuroprotective agents being studied for Alzheimer's?

Antioxidants, such as vitamins C and E, are studied for their potential to protect brain cells from oxidative stress, which contributes to Alzheimer's. Neuroprotective agents like curcumin (from turmeric) are also under investigation for their potential to reduce plaque formation and inflammation. While early studies are promising, more robust clinical trials are needed to confirm their effectiveness in Alzheimer's patients.

Q46: What are the potential benefits and risks of hormone replacement therapy for Alzheimer's?

Hormone replacement therapy (HRT) has been studied for Alzheimer's, especially in postmenopausal women. Estrogen protects brain cells, and some studies suggest it may lower Alzheimer's risk. However, HRT also carries risks, such as an increased chance of cardiovascular disease and certain cancers. Current guidelines advise against using HRT solely for Alzheimer's prevention.

Q47: How does the ketogenic diet impact Alzheimer's symptoms?

The ketogenic diet, high in fats and low in carbohydrates, may provide an alternative energy source for the brain in the form of ketones. Early studies indicate it could improve cognitive function by reducing insulin resistance and promoting brain energy metabolism. However, the diet is restrictive, and the long-term safety and effectiveness in Alzheimer's patients are still under research.

Q48: What is the role of social and recreational activities in Alzheimer's therapy?

Social and recreational activities help maintain cognitive and emotional health by stimulating the brain, reducing isolation, and improving mood. Activities like group outings, games, or creative projects engage patients and may slow cognitive decline. Social interaction is crucial, as it can reduce loneliness and enhance quality of life.

Q49: How do meditation and mindfulness therapy benefit Alzheimer's patients?

Meditation and mindfulness therapy can help reduce stress and improve mood in Alzheimer's patients, offering moments of calm and focus. Regular mindfulness practice may also improve sleep and reduce anxiety, helping both patients and caregivers manage the challenges of daily life with the disease.

Q50: How can family therapy support caregivers and patients with Alzheimer's?

Family therapy offers emotional support, guidance, and resources to both Alzheimer's patients and their caregivers. It can help families cope with the stress and emotional challenges of caregiving, promote effective communication, and ensure that everyone understands the needs and progression of the disease. Family therapy also fosters a supportive environment, improving overall care quality.

Q51: How are digital health tools, like apps, used to support Alzheimer's management?

Digital health tools and apps increasingly remind patients to take medications, track symptoms, and provide cognitive exercises. These tools often help caregivers monitor and assist with daily routines, offering structured support for both patients and caregivers. They also provide resources for education, memory aids, and social engagement.

Q52: What are the latest breakthroughs in Alzheimer's disease vaccine research?

Recent advancements focus on developing vaccines that target beta-amyloid plaques and tau proteins, two hallmarks of Alzheimer's pathology. Some experimental vaccines stimulate the immune system to clear these abnormal proteins. While still in clinical trial phases, several vaccine candidates have shown promise, and ongoing research aims to bring effective vaccines closer to clinical use.

Q53: How do structured routines and daily planning help manage Alzheimer's symptoms?

Structured routines and daily planning reduce confusion and provide stability for Alzheimer's patients. Consistent schedules make it easier for patients to remember daily tasks and reduce anxiety associated with uncertainty. Routines are especially helpful in managing behavioral symptoms, as predictable activities create a calm environment.

Q54: What safety measures should be considered in Alzheimer's care at home?

Safety measures include removing tripping hazards, installing door locks to prevent wandering, using non slip mats, and securing dangerous items like cleaning supplies or medications. Clear labeling, monitoring devices, and ensuring adequate lighting are also essential. These modifications help create a safer environment and reduce the risk of accidents.

Q55: How can environmental modifications reduce stress for Alzheimer's patients?

Environmental modifications, like calming lighting, removing clutter, adding visual cues, and maintaining a familiar layout, reduce stress by creating a comfortable and accessible space. Personalized decorations, such as photos, can evoke positive emotions and enhance a sense of belonging, creating a reassuring environment for Alzheimer's patients.

Q56: What are the best practices in palliative care for advanced Alzheimer's patients?

Palliative care focuses on comfort, pain management, and emotional support, prioritizing quality of life over aggressive treatment. Best practices include:

- Managing symptoms like pain.
- Providing compassionate communication.
- Involving family members in care decisions.
- Addressing emotional and spiritual needs.

Palliative care is centered on dignity, respect, and comfort.

Q57: What is the potential of CRISPR technology in treating Alzheimer's disease?

CRISPR gene-editing technology is being researched as a potential approach to alter genes linked to Alzheimer's risk, such as APOE-e4. Scientists hope to reduce risk or slow progression by editing these genes, although CRISPR research in Alzheimer's is still in its early stages. Ethical concerns and technical challenges must be addressed before this approach can be considered for clinical use.

Q58: How are sleep therapies used to improve outcomes for Alzheimer's patients?

Sleep therapies, including cognitive behavioral therapy for insomnia (CBT-I), melatonin supplementation, and sleep hygiene education, help improve sleep patterns in Alzheimer's patients. Proper sleep can improve mood, cognitive functioning, and overall well-being, making sleep therapy an essential aspect of Alzheimer's care.

Q59: What role does hydration play in Alzheimer's disease symptom management?

Hydration is crucial, as dehydration can exacerbate confusion, irritability, and cognitive symptoms in Alzheimer's patients. Maintaining adequate hydration helps prevent urinary tract infections and other health issues common in these patients, supporting overall physical and mental health.

Q60: What are the latest trends in Alzheimer's rehabilitation programs?

Newer rehabilitation programs combine cognitive training, physical exercise, and social engagement to slow symptom progression and maintain quality of life. Virtual reality, digital cognitive games, and adaptive technologies are emerging trends that enable patients to interact in therapeutic environments. These programs aim to keep patients mentally and physically engaged.

Q61: When were the first drugs approved to treat Alzheimer's, and what were their primary functions?

The first drugs approved to treat Alzheimer's disease were cholinesterase inhibitors, with the earliest approval being tacrine (Cognex) in 1993. Tacrine prevents the breakdown of acetylcholine, a neurotransmitter essential for memory and learning, which is often depleted in Alzheimer's patients. By increasing acetylcholine levels in the brain, tacrine helped improve cognitive function in some patients, though it had limited effectiveness and was associated with significant side effects, including liver toxicity.

Following tacrine, other cholinesterase inhibitors with fewer side effects were developed and approved. These include donepezil (Aricept) in 1996, rivastigmine (Exelon) in 2000, and galantamine (Razadyne) in 2001. These medications also aim to increase acetylcholine levels to support cognitive function, memory, and daily activities. Though not cures, they can help temporarily slow the worsening of symptoms in the mild to moderate stages of Alzheimer's disease.

In 2003, the first NMDA receptor antagonist, memantine (Namenda), was approved to treat moderate to severe Alzheimer's. Memantine works differently than cholinesterase inhibitors; it regulates glutamate, another neurotransmitter involved in learning and memory. Excessive glutamate activity can damage cells, so memantine helps protect brain cells by blocking abnormal glutamate activity. Together, these drugs remain foundational in managing Alzheimer's symptoms, though they do not halt disease progression.

Q62: Is there any evidence to suggest that marijuana use can slow the progression of Alzheimer's disease?

Currently, there is no conclusive evidence that marijuana use can slow the progression of Alzheimer's disease. However, early studies suggest that specific components of marijuana, particularly CBD (cannabidiol), may have neuroprotective properties that could be beneficial. CBD has shown anti-inflammatory and antioxidant effects in laboratory and animal studies, which may help protect brain cells from damage. Since inflammation and oxidative stress contribute to the progression of Alzheimer's, some researchers hypothesize that CBD could play a supportive role in reducing these harmful processes. However, these findings are still preliminary, and human trials are needed to determine whether CBD has a significant impact on Alzheimer's progression.

Additionally, some studies indicate that THC and other cannabinoids may help reduce the buildup of beta-amyloid plaques, one of the pathological markers of Alzheimer's. THC appeared to lower amyloid production in lab models and potentially reduce plaque accumulation. However, these effects have not been consistently replicated in human studies, and the cognitive side effects of THC—such as impairing short-term memory and increasing confusion—raise concerns about its practical use in Alzheimer's patients. Ultimately, while cannabinoids show promise in preclinical studies, there is not yet enough clinical evidence to support the idea that marijuana can slow Alzheimer's disease progression in humans. Rigorous, large-scale trials are needed to confirm or refute these early findings.

Q63: What effects does marijuana have on Alzheimer's disease symptoms?

Marijuana, particularly compounds like CBD (cannabidiol) and THC (tetrahydrocannabinol), may help alleviate specific symptoms of Alzheimer's disease, although the effects vary based on the type and dosage used. CBD, for instance, has shown promise in managing anxiety, agitation, and depression, which are common in Alzheimer's patients. Known for its calming effects, CBD may help reduce stress and anxiety without causing the " high " associated with THC, offering a potential way to ease mood-related symptoms. Additionally, CBD's anti-inflammatory and antioxidant properties may provide neuroprotective effects, which could theoretically help slow the progression of symptoms, though more research is needed in this area.

THC, on the other hand, may help with pain management, appetite stimulation, and sleep, which can be beneficial for Alzheimer's patients who experience chronic discomfort, weight loss, or insomnia. THC has mild sedative effects that may promote sleep and calm agitation, but it also carries risks of cognitive side effects, especially at higher doses. High THC levels can worsen confusion, impair short-term memory, and increase the risk of paranoia in some individuals, which may exacerbate Alzheimer's symptoms. For these reasons, any potential therapeutic use of marijuana for Alzheimer's symptoms requires careful management and dosing, ideally with medical oversight, to mitigate risks and maximize benefits.

Q64: What effects do psychedelics have on brain health and Alzheimer's disease?

Psychedelics such as psilocybin, LSD, and MDMA are being studied for their potential effects on brain health, particularly neuroplasticity, mood regulation, and inflammation—all factors relevant to Alzheimer's disease. These substances interact with serotonin receptors in the brain, particularly the 5-HT2A receptor, which is linked to cognition, mood, and perception. By activating these receptors, psychedelics appear to promote neuroplasticity, or the brain's ability to form and reorganize synaptic connections. This enhanced plasticity may help the brain maintain cognitive function longer or compensate for early neurodegenerative damage, although this hypothesis is still under investigation.

In addition to promoting neuroplasticity, psychedelics may have anti-inflammatory effects that could benefit Alzheimer's patients. Chronic brain inflammation contributes to the progression of Alzheimer's, and early research suggests that psychedelics might reduce inflammation by modulating immune responses in the brain. MDMA has shown promise in addressing anxiety, depression, and social withdrawal, symptoms often experienced by Alzheimer's patients, by promoting serotonin release and enhancing feelings of connectedness and calm. While these effects suggest potential therapeutic benefits for symptom management, there is currently no evidence that psychedelics directly prevent or reverse Alzheimer's disease. Clinical trials are still in the early stages, and more research is needed to assess both the safety and long-term efficacy of psychedelics for Alzheimer's and other neurodegenerative conditions.

Q65: How does spending time in nature affect people with Alzheimer's disease?

Spending time in nature has shown positive effects on people with Alzheimer's disease. Research suggests that being outdoors can help reduce agitation, improve mood, and enhance cognitive function, even if only temporarily. Nature offers a calming environment that can reduce stress and anxiety, common symptoms of Alzheimer's. Additionally, exposure to natural settings promotes physical health by encouraging mobility and physical activity.

The sights, sounds, and smells of nature stimulate sensory responses, which may help trigger memories or feelings of familiarity. Furthermore, the peaceful, less stimulating environment of natural spaces may be less overwhelming than busy indoor or urban settings, making it easier for people with Alzheimer's to relax and feel at ease.

These benefits are thought to be linked to nature's ability to reduce cognitive load and provide a sense of security and comfort, both of which are essential for individuals with memory impairments.

Q66: How do outdoor or nature-based therapy programs compare to traditional indoor therapies in terms of improving the quality of life for Alzheimer's patients?

Outdoor or nature-based therapy programs often offer distinct benefits over traditional indoor therapies when it comes to improving the quality of life for individuals with Alzheimer's disease. Nature-based therapies, such as horticultural therapy, outdoor walks, and animal-assisted activities, provide a multisensory experience that stimulates memory, enhances mood, and reduces anxiety. These activities promote physical exercise, which is beneficial for overall health and cognitive function, while also encouraging social interaction in a less stimulating, more relaxed environment than typical indoor settings. The calming effect of nature, coupled with sensory engagement (e.g., the sights, sounds, and smells of plants and wildlife), can be particularly effective in reducing agitation and improving emotional well-being.

In contrast, traditional indoor therapies like cognitive stimulation therapy or memory exercises focus more on structured mental tasks, which may not provide the same level of sensory or emotional relief. While these therapies are crucial for cognitive engagement, outdoor environments often offer more immediate mood-boosting effects. Indoor therapy sessions can also be less motivating for individuals with Alzheimer's, who may feel confined or overwhelmed by artificial settings. Overall, nature-based programs offer a holistic approach to care, addressing both physical and emotional needs, and often lead to better engagement and quality of life than traditional indoor therapies.

–

In conclusion, while Alzheimer's disease remains without a cure, the array of treatments and interventions available today offers hope for improving the lives of those affected by the disease. From FDA-approved medications that manage cognitive and behavioral symptoms to emerging therapies targeting the underlying biological causes, significant progress has been made. Additionally, non-pharmacological approaches, including cognitive therapies, physical activity, and social engagement, provide essential support for mental and emotional well-being. Lifestyle changes and caregiver education further enhance the quality of care and create a supportive environment for both patients and families. As research continues to advance, these combined efforts lay a strong foundation for managing Alzheimer's disease and improving the lives of those living with it.

Bibliography

1. **Haeberlein, S. F., Cummings, J., Cohen, S., McCallum, T., & Anderson, R. (2019).** *Current and emerging treatments for Alzheimer's disease. J Neurol , 266(1), 187– 201.*
2. **van Dyck, C. H. (2018).** *Anti-amyloid-beta monoclonal antibodies for Alzheimer's disease: Past, present, and future. Biological Psychiatry, 83(4), 311–31 9.*
3. **Sperling, R. A., Aisen, P. S., Beckett, L. A., Bennett, D. A., Craft, S., Froelich, L., et al. (2011).** *Toward defining the preclinical stages of Alzheimer's disease: Recommendations from the National Institute on Aging-Alzheimer's Association workgroups on diagnostic guidelines for Alzheimer's disease. Alzheimer's & Dement ia, 7(3), 280–2 92.*

Living with Alzheimer's Disease

CHAPTER 5

Living with Alzheimer's disease involves navigating the physical, cognitive, and emotional challenges that come with the condition while striving to maintain a sense of dignity and purpose. This journey requires practical strategies to manage daily life, such as establishing consistent routines, ensuring a safe and supportive home environment, and fostering as much independence as possible during the early stages. Emotional well-being is equally vital, with techniques for coping with anxiety, depression, and feelings of isolation playing a key role in enhancing quality of life.

For caregivers, understanding how to communicate effectively, manage behavioral changes, and build support networks is essential to providing compassionate care while preventing burnout. Planning is another critical aspect, including legal and financial preparations and discussions about long-term care options. Community programs, support groups, and innovative technologies also offer valuable resources to help patients and their loved ones navigate this complex journey. By addressing the practical, emotional, and social dimensions of the disease, these strategies provide a foundation for resilience, connection, and hope in the face of Alzheimer's.

–

Q1: What are the environmental modifications for individuals with Alzheimer's disease?

Environmental modifications can significantly enhance safety and comfort for individuals with Alzheimer's disease. Examples include reducing clutter, improving lighting, installing handrails and grab bars, and using contrasting colors to improve visibility. Creating a familiar, structured environment with clear signage and visual cues helps reduce confusion and promotes independence. It is also important to regularly assess and adapt the environment as the disease progresses to ensure continued support and safety.

▲ **FIGURE 5.1.**
An Elderly Man Using His Mobile Phone while Holding a Credit Card

Q2: What lifestyle adjustments can help individuals with Alzheimer's disease live more independently?

Lifestyle adjustments can empower individuals with Alzheimer's disease to maintain independence and quality of life. These adjustments include establishing routines, using memory aids such as calendars and reminders, simplifying tasks, and organizing living spaces for safety and ease of navigation. Engaging in meaningful activities and social interactions also promotes well-being and cognitive function.

The literature related to lifestyle management options for Alzheimer's disease (AD) is vast. This section will draw from an article titled "Current and Future Treatments in Alzheimer's Disease" by Alireza Atri, which provides a comprehensive overview of the multifactorial approach to managing AD. It explores both pharmacological and non-pharmacological strategies aimed at mitigating symptoms and slowing AD progression. The author discusses current treatments, the importance of caregiver involvement, and emerging research on disease-modifying therapies. This paper highlights the critical role of lifestyle interventions alongside medication to support individuals with AD in maintaining independence and quality of life.

Lifestyle adjustments that help individuals with AD maintain independence can be highly beneficial. The article emphasizes a multipronged approach that combines pharmacological treatments with non-pharmacological strategies tailored to the specific needs of the patient and caregiver. One key non-pharmacological intervention is psychoeducation, which involves educating both the patient and their caregivers about the disease, its progression, and its effects on cognition and behavior. This ongoing process enables families and caregivers to better understand the condition and adjust care strategies accordingly.

Another vital lifestyle adjustment is environmental modification. Simplifying the living space, reducing clutter, and establishing consistent routines can help manage symptoms such as confusion and anxiety. A calm and predictable environment reduces disorientation and enhances the patient's ability to perform daily tasks independently.

Additionally, implementing behavioral approaches can address neuropsychiatric symptoms commonly seen in AD, such as agitation or aggression. Techniques like calmly interacting with the patient, redirecting them to enjoyable activities, and using positive reinforcement are beneficial strategies. These methods aim to reduce distress and promote a more peaceful environment, making it easier for the patient to engage in daily activities. Support networks for both the patient and the caregivers are also essential. Caregivers often experience high stress levels, and establishing a network of support—whether through respite care or community resources—helps maintain the well-being of both parties. Finally, monitoring health and safety risks is crucial for patient independence. Regular assessments for fall risks, medication safety, and cognitive decline allow for timely interventions that prevent accidents or dangerous situations, such as cooking or driving hazards.

When integrated with appropriate pharmacological treatments, these non-pharmacological interventions create a holistic care plan that can significantly improve the quality of life and independence of individuals with AD.

Q3: How can caregivers support individuals with Alzheimer's disease in daily activities?

Caregivers play a vital role in supporting individuals with Alzheimer's disease in daily activities. Strategies include:

- Breaking tasks into manageable steps.
- Providing clear instructions and prompts.

- Helping as needed while preserving autonomy.
- Maintaining patience and flexibility.

Caregivers can also create a supportive environment that minimizes distractions and reduces stress for the individual with Alzheimer's.

Q4: What are the challenges and benefits of engaging in social activities for individuals with Alzheimer's disease?

Engaging in social activities offers numerous benefits for individuals with Alzheimer's disease, including cognitive stimulation, emotional support, and enhanced quality of life. Social interactions help maintain communication skills and cognitive function by providing mental stimulation and opportunities for engagement. Additionally, socializing reduces loneliness and isolation, fostering a sense of belonging and emotional well-being. However, there are challenges associated with social activities for individuals with Alzheimer's, such as difficulties in communication and expressing thoughts, which may lead to frustration or withdrawal. Sensory overload in noisy or crowded environments can also cause discomfort or agitation. Careful planning and adaptation of activities to suit the individual's abilities and preferences, along with providing support and reassurance, can help overcome these challenges and maximize the benefits of social engagement.

The article "Leisure Connections: Benefits and Challenges of Participating in a Social Recreation Group for People with Early Dementia" by Alison Phinney and Elaine M. Moody provides a detailed examination of the benefits and challenges faced by individuals with early dementia as they engage in social activities within a community-based setting. Phinney and Moody's study focuses on the Leisure Connections program, designed specifically for people with early dementia to foster social interaction and support through structured group activities. Using qualitative methods, including interviews and participant observation, the study highlights the participants' perspectives on how these social activities impact their quality of life, social relationships, and overall well-being.

One of the study's key findings is that participating in social activities provides significant emotional and social benefits for individuals with early dementia. Participants frequently noted the sense of camaraderie and acceptance within the group, which created a unique space to engage with others who understood their challenges. The group environment allowed participants to share a common understanding of cognitive difficulties, making them feel comfortable, safe, and free from judgment. This sense of belonging was deeply valued, as many participants felt isolated in their daily lives, often due to the stigma surrounding dementia. By participating in a group with others who had similar experiences, they could express themselves more freely and enjoy moments of laughter and joy without fear of embarrassment. This social acceptance promoted a positive atmosphere where members felt appreciated and supported, reinforcing the therapeutic potential of social engagement for people with dementia.

The emotional benefits of the program extended beyond the social interactions themselves. Participants reported increased feelings of purpose and self-worth due to their involvement. Activities were intentionally designed to allow participants to contribute to the group by sharing life stories, assisting each other with tasks, or even providing comedic relief. Such opportunities for contribution helped combat feelings of dependency and helplessness, which are common among individuals with dementia. Phinney and Moody emphasize that these moments of contribution and responsibility are integral in fostering a sense of empowerment and self-respect, as participants felt valued for their abilities rather than defined by their limitations. This stands in contrast to more medically oriented settings, where individuals with dementia are often passive recipients of care. Instead, the social recreation group encouraged active participation and autonomy, enhancing self-esteem and a sense of identity.

The study also sheds light on challenges associated with facilitating social activities for individuals with dementia, particularly as the disease progresses. Phinney and Moody note that while the group structure benefited early-stage participants, it became increasingly difficult to maintain as some individuals experienced cognitive decline. Staff and volunteers observed that managing participants' needs, given their varying cognitive abilities, required a careful balance, as the group's social cohesion depended on participants' ability to interact meaningfully. Additionally, family members and staff faced the emotional challenge of recognizing when individuals needed to transition to more intensive support programs as their abilities declined. This highlights a limitation of community-based programs, as they may not have the resources or structure to accommodate advanced dementia, ultimately necessitating a move to a different care setting. Phinney and Moody suggest that this issue could be mitigated by viewing programs like Leisure Connections as a "stepping-stone" to more structured day programs, allowing participants to gradually transition as needed.

In conclusion, Phinney and Moody's study underscores the multifaceted benefits and inherent challenges of engaging individuals with dementia in social activities. The emotional connections and sense of purpose fostered within the Leisure Connections group illustrate the profound impact of meaningful social engagement on the well-being of individuals with dementia. At the same time, the study highlights the complexities involved in maintaining such programs, particularly as participants' cognitive abilities evolve. These findings contribute valuable insights into the design and implementation of community-based dementia programs, emphasizing the importance of fostering environments that support social inclusion and empowerment while addressing participants' evolving needs as dementia progresses.

Q5: How can technology assist individuals with Alzheimer's disease and their caregivers?

Technology offers valuable tools for individuals with Alzheimer's disease and their caregivers, including memory aids, safety devices, communication apps, and remote monitoring systems. Wearable devices, GPS trackers, and smart home technologies can enhance safety and security, while telemedicine and virtual support groups provide convenient access to healthcare and social support. Careful selection and customization of technology solutions can help individuals with Alzheimer's maintain independence and improve their quality of life.

Q6: What are the legal and financial considerations for individuals with Alzheimer's and their families?

Legal and financial planning are crucial for individuals with Alzheimer's disease and their families to ensure that their wishes are respected, and their needs are met as the disease progresses.

Key legal considerations include establishing powers of attorney for healthcare and finances, creating a living will, and setting up guardianship if necessary. These documents allow designated decision-makers to act on behalf of the individual when they can no longer make decisions themselves.

Financial planning involves organizing assets, creating trusts, and accessing resources such as government benefits and long-term care insurance to cover care costs. Consulting with elder law attorneys and financial professionals early in the disease process can help families navigate these complex issues, protect the individual's interests, and make informed decisions for the future.

Q7: How can individuals with Alzheimer's disease and their families prepare for end-of-life care?

End-of-life care planning involves discussing preferences for medical treatment, palliative care, and hospice services with individuals with Alzheimer's disease and their families. Advance care directives, such as living wills and do-not-resuscitate orders, should be documented and communicated to healthcare providers. Families may also consider emotional and spiritual support, funeral arrangements, and legacy planning to ensure a dignified and peaceful end-of-life experience for their loved ones with Alzheimer's.

The article "Advance Care Planning in Dementia: Recommendations for Healthcare Professionals" offers valuable guidance on preparing individuals with Alzheimer's and their families for end-of-life care. It emphasizes the importance of Advance Care Planning (ACP), a continuous process of reflection and dialogue involving the patient, their family, and healthcare professionals to discuss preferences for future care, including end-of-life decisions.

To ensure that individuals with Alzheimer's and their families are well-prepared for end-of-life care, it is crucial to initiate ACP conversations as early as possible, ideally in the early stages of dementia. This allows patients to express their values and preferences while they still have the mental capacity to do so. Early initiation ensures that their wishes regarding future treatments and care, including life-prolonging interventions or comfort-focused care, are documented.

Another vital aspect is involving family members in these conversations early on. Since individuals with Alzheimer's gradually lose decision-making abilities, family members or close friends are often required to act as surrogate decision-makers. Engaging them early helps align their understanding of the patient's wishes and provides emotional support.

Documentation is a critical component of ACP. Families and healthcare professionals should ensure that the individual's preferences are recorded, such as through an advance directive or by appointing a legal representative. This documentation serves as a guide for decision-making when the patient can no longer communicate their wishes.

Lastly, ACP should include regular re-evaluations as the disease progresses. Decisions about the patient's care may change significantly as their health deteriorates. Therefore, it is essential to revisit and update care goals and preferences to ensure they continue to reflect the patient's evolving condition and wishes.

By engaging in advance care planning, individuals with Alzheimer's and their families can ensure that their end-of-life care aligns with their values, reducing stress and uncertainty during the later stages of the disease.

Q8: What support services are available for individuals with Alzheimer's disease and their caregivers?

Various support services are available to assist individuals with Alzheimer's disease and their caregivers throughout the disease journey. These include Alzheimer's Association helplines, support groups, respite care programs, adult day centers, and home care services. Memory cafes, educational workshops, and online resources provide information, connection, and practical assistance for individuals with Alzheimer's and their caregivers. Accessing these support services can help alleviate caregiver burden, enhance quality of life, and promote well-being for everyone affected by Alzheimer's disease.

Q9: What is the role of support groups for Alzheimer's disease?

Support groups offer valuable resources for individuals with Alzheimer's disease and their caregivers. These groups provide a safe space to share experiences, seek advice, and receive emotional support from others facing similar challenges. Support groups can help reduce feelings of isolation, offer practical coping strategies, and enhance the overall well-being of participants. Many organizations and communities provide both in-person and online support group options.

Q10: How can individuals with Alzheimer's disease and their families advocate for better care and support services?

Advocacy efforts play a crucial role in improving care and support services for individuals with Alzheimer's disease and their families. This can involve raising awareness, sharing personal experiences, and contacting policymakers to advocate for increased funding, research initiatives, and dementia-friendly policies. Participating in clinical trials, joining advocacy organizations, and speaking out about the needs of the Alzheimer's community can drive positive change and improve outcomes for future generations.

Q11: What lifestyle adjustments can help individuals with Alzheimer's disease live more independently?

Lifestyle adjustments can empower individuals with Alzheimer's disease to maintain independence and quality of life. These include establishing daily routines to provide structure and reduce confusion, using memory aids such as calendars and reminders for important tasks, simplifying tasks into manageable steps, and organizing living spaces for safety and ease of navigation. Engaging in meaningful activities and social interactions can also promote well-being and cognitive function, supporting overall independence despite the challenges of the disease.

Q12: How can caregivers support individuals with Alzheimer's disease in daily activities?

Caregivers play a crucial role in supporting individuals with Alzheimer's disease in daily activities. Strategies include breaking tasks into simple steps and providing clear, step-by-step instructions. Visual cues and prompts can help individuals understand and complete tasks more independently. Caregivers should maintain patience, offer reassurance, and allow ample task time. Preserving the person's autonomy is essential by encouraging them to do as much as possible independently while assisting when needed. Creating a calm and supportive environment minimizes stress and frustration, enhancing the individual's ability to engage in daily activities.

Q13: What are the challenges and benefits of engaging in social activities for individuals with Alzheimer's disease?

Engaging in social activities offers numerous benefits for individuals with Alzheimer's, including cognitive stimulation, emotional support, and enhanced quality of life. Social interactions help maintain communication skills and cognitive function by providing mental stimulation and opportunities for engagement. Additionally, socializing reduces loneliness and isolation, fostering a sense of belonging and emotional well-being. However, there are challenges associated with social activities for individuals with Alzheimer's disease. These include difficulties in communication and expressing thoughts, leading to frustration or withdrawal from social interactions. Sensory overload in noisy or crowded environments may also cause discomfort or agitation. Careful planning and adaptation of activities to suit the individual's abilities and preferences, along with providing support and reassurance, can help overcome these challenges and maximize the benefits of social engagement.

Q14: How can technology assist individuals with Alzheimer's disease and their caregivers?

Technology plays a pivotal role in assisting individuals with Alzheimer's disease and their caregivers by providing practical solutions to everyday challenges. For individuals with Alzheimer's, technology offers memory aids such as reminder apps and electronic calendars to help them stay organized and maintain routines independently. Safety devices like GPS trackers and smart home systems ensure security and prevent wandering. Communication apps facilitate easy interaction with loved ones, helping to overcome communication barriers often associated with the disease. For caregivers, technology provides support tools, including caregiver apps and online resources, offering information, guidance, and connections to support networks. Telemedicine enables remote healthcare access, reducing the burden of frequent medical visits. Overall, technology enhances the quality of life, safety, and caregiving efficiency for both individuals with Alzheimer's disease and their caregivers.

The article "Behavioral Interventions for Alzheimer's Management Using Technology: Home-Based Monitoring" explores how technology can support individuals with Alzheimer's disease (AD) and their caregivers in home settings. It provides a comprehensive review of current technological tools used to monitor behavioral symptoms and improve the quality of life for both patients and caregivers.

One of the primary ways technology assists individuals with Alzheimer's is through home-based monitoring systems. These systems use sensors and video recordings to track the patient's movements, behaviors, and activities of daily living (ADLs). For example, wearable and in-home environmental sensors can detect agitation and aggression in patients with dementia, providing real-time data that caregivers and healthcare professionals can use to respond proactively. These technologies help monitor safety and detect early signs of behavioral disturbances, allowing for timely interventions.

Additionally, telehealth systems combined with video recording can support caregivers by providing remote expert analysis of caregiving techniques and patient behaviors. This not only improves caregivers' competence but also reduces their depression and stress. These systems allow caregivers to capture videos of the patient's behavior, which can be reviewed by healthcare professionals, who then provide guidance on managing symptoms more effectively.

Innovative health tools and assistive devices also enhance the patient's ability to engage in daily tasks independently. For instance, prompting systems guide individuals with dementia through tasks like handwashing, using cameras and audio cues to reduce the caregiver's workload. These systems help maintain a routine and reduce the need for constant supervision.

In summary, technology offers a wide range of solutions to assist individuals with Alzheimer's disease, from monitoring safety and managing behavioral symptoms to reducing caregiver burden and enhancing patient autonomy in daily activities. When used effectively, these interventions can significantly improve the patient's quality of life and the caregiver's well-being.

Q15: What are the legal and financial considerations for individuals with Alzheimer's disease and their families?

Legal and financial planning is essential for individuals with Alzheimer's disease and their families. It involves setting up advance directives like power of attorney for healthcare and finances, creating or updating wills, managing finances and accessing benefits, considering legal guardianship if needed, and planning long-term care costs. Early consultation with legal and financial professionals helps ensure that necessary protections are in place and enables informed decision-making to support the well-being and future care of those affected by Alzheimer's disease.

Q16: How can individuals with Alzheimer's disease and their families prepare for end-of-life care?

Preparing for end-of-life care for individuals with Alzheimer's disease and their families involves discussing and documenting preferences for medical care and decision-making through advance directives, such as living wills and healthcare proxies. It also includes ensuring that legal and financial affairs are in order, addressing emotional and spiritual needs, and making practical arrangements for funeral or memorial plans. Open communication among family members and healthcare providers is crucial to ensure that the individual's wishes are understood and respected, ultimately providing a supportive and dignified end-of-life experience.

Q17: What support services are available for individuals with Alzheimer's disease and their caregivers?

Several support services are available to assist individuals with Alzheimer's disease and their caregivers. These include Alzheimer's associations, which offer helplines and local chapters providing information and support groups. Respite care programs offer temporary relief for caregivers, while adult day centers provide structured activities and supervision. Home care services deliver professional assistance at home, and memory cafes offer social interaction in a supportive environment. Educational workshops and webinars also provide valuable resources for understanding Alzheimer's and developing caregiving skills. These services aim to alleviate caregiver stress, enhance quality of life, and provide practical and emotional support throughout the progression of Alzheimer's disease.

However, despite the availability of these support services, several factors influence whether individuals and caregivers choose to access them. The article "Barriers and Facilitators of Using Dementia Community Support Services" provides insight into the challenges and enablers for Alzheimer's patients and their caregivers when accessing support services. Despite the services offered, the study highlights a range of barriers that prevent effective utilization and identifies factors that can facilitate service use.

One of the most significant barriers is a lack of awareness. Many caregivers are not fully informed about the services provided by Alzheimer's organizations. Caregivers often reported that the insufficient promotion and visibility of available services hinder their ability to seek help. Additionally, personal barriers, such as grief, guilt, and a desire for privacy, play a significant role in avoiding service use. Caregivers frequently feel guilty for seeking external help, believing they should handle everything independently. They may also feel embarrassed, or fear being misunderstood, contributing to their reluctance to reach out for support.

Conversely, the study identifies several facilitators that encourage the use of support services. Empathetic staff members who build strong, trusting relationships with caregivers and individuals with Alzheimer's are key facilitators. Caregivers appreciate having a consistent support person who understands their emotional needs and helps normalize their experiences. Proactive contact from the support organization is another facilitator. Many caregivers admitted they would not have sought help on their own, but regular, proactive check-ins by the organization encouraged them to engage with services. Support groups are also highlighted as valuable, providing caregivers with a sense of community and reducing feelings of isolation. Finally, providing information and education helps caregivers feel more confident and informed about managing Alzheimer's disease, making them more likely to use services.

In summary, while support services are available for individuals with Alzheimer's and their caregivers, barriers such as lack of awareness, emotional challenges, and a sense of burden can prevent their use. However, facilitators such as empathetic staff, proactive outreach, and educational resources can help overcome these obstacles and ensure caregivers receive the support they need.

Q18: How can individuals with Alzheimer's disease and their families advocate for better care and support services?

Individuals with Alzheimer's disease and their families can advocate for better care and support services by raising awareness about the disease through personal stories and community engagement. Joining advocacy organizations enables them to support initiatives aimed at improving services and policies. Contacting policymakers to advocate for increased funding for research and better access to healthcare is crucial. Participating in research studies and clinical trials can contribute to advancements in treatment and care. Sharing experiences through media and community events helps educate the public and promote understanding of the challenges faced by those with Alzheimer's. By advocating collectively, individuals can influence policies, create dementia-friendly communities, and improve the overall quality of care and support available for those affected by Alzheimer's disease.

Q19: How can individuals with Alzheimer's disease maintain a sense of purpose and fulfillment?

Individuals with Alzheimer's can maintain a sense of purpose by engaging in activities they enjoy or once enjoyed, such as hobbies, volunteering, or contributing to family life. Encouraging participation in meaningful activities, like art, gardening, or helping with household tasks, can enhance their sense of self-worth and provide structure to their day.

Q20: What role does routine play in the daily life of individuals with Alzheimer's disease?

A routine provides structure, security, and predictability, which are essential for individuals with Alzheimer's. Having a set routine can help reduce anxiety, confusion, and agitation, making it easier for them to remember daily tasks and feel more in control of their environment. Routines also assist caregivers in managing the patient's care efficiently.

Q21: How can caregivers manage stress and prevent burnout when caring for someone with Alzheimer's?

Caregivers can manage stress by prioritizing self-care, seeking emotional support from family, friends, or support groups, and taking regular breaks. It's important to establish a routine that includes time for relaxation, hobbies, and physical activity.

Additionally, professional help, such as respite care, can alleviate stress and prevent burnout.

Q22: How can family members communicate effectively with a loved one with Alzheimer's?

Effective communication with Alzheimer's patients requires patience, simple language, and maintaining eye contact. It's important to avoid complex questions, use gestures or visual cues, and ensure a calm, quiet environment. Listening attentively and offering reassurance can help patients feel understood and supported.

Q23: What dietary adjustments may help improve the quality of life for those with Alzheimer's disease?

A diet rich in fruits, vegetables, whole grains, and healthy fats—such as the Mediterranean diet—may help support brain health. Omega-3 fatty acids, antioxidants, and B vitamins have also been linked to improved cognitive function. Avoiding processed foods and ensuring regular meals can contribute to maintaining overall health.

Q24: How can exercise benefit individuals with Alzheimer's disease and their caregivers?

Exercise can improve mood, boost cognitive function, and enhance physical health in Alzheimer's patients. It helps reduce anxiety, depression, and agitation while maintaining mobility and strength. For caregivers, physical activity can alleviate stress and improve overall well-being, supporting them in managing the demands of caregiving.

Q25: What are some effective relaxation techniques for people with Alzheimer's disease?

Techniques such as deep breathing, progressive muscle relaxation, and guided meditation can help reduce anxiety and promote calm. Regular relaxation exercises, like listening to soothing music or engaging in gentle stretching or yoga, can also enhance emotional well-being and alleviate restlessness.

Q26: How can Alzheimer's patients express their wishes for future care?

Alzheimer's patients can express their wishes through advance directives, living wills, or durable power of attorney documents.

Family members should encourage early discussions about preferences for medical care, living arrangements, and end-of-life decisions, helping ensure that the patient's wishes are respected as the disease progresses.

Q27: How can caregivers create a safe home environment for someone with Alzheimer's?

Caregivers can create a safe environment by removing tripping hazards, installing grab bars, and using non slip rugs. Dangerous items, such as cleaning products and sharp objects, should be secured, and labels can be used to aid memory. Additionally, installing adequate lighting, especially in hallways and bathrooms, helps prevent accidents.

Q28: How can pets or service animals improve life for people with Alzheimer's disease?

Pets and service animals can offer companionship, reduce anxiety, and improve mood in individuals with Alzheimer's. They provide emotional support, help alleviate feelings of loneliness, and can even assist with daily tasks. Additionally, the presence of pets can encourage physical activity and contribute to a sense of routine and purpose.

Q29: How can caregivers and family members handle changes in personality or behavior?

Caregivers can manage personality and behavior changes by remaining patient, using a calm tone, and redirecting conversations or actions when needed. It's important to stay flexible and remember that these changes are a part of the disease, not a reflection of the person's character. Consulting with a healthcare provider can help address mood swings or aggression.

Q30: How can art and creative activities improve the quality of life for individuals with Alzheimer's?

Art and creative activities provide individuals with Alzheimer's an opportunity to express themselves in ways that words may no longer capture. These activities can improve mood, reduce anxiety, and stimulate cognitive function. Whether through painting, drawing, music, or dance, creative expression offers both emotional and mental stimulation.

Q31: What are the best ways to encourage physical activity for individuals with Alzheimer's?

Physical activity can be encouraged through enjoyable, low-impact activities like walking, swimming, or simple exercises. Caregivers can participate in these activities to make them more social and motivating. Structured programs, such as chair exercises or dance classes, can also safely engage individuals with Alzheimer's in physical activity.

Q32: How can religious or spiritual practices support individuals with Alzheimer's and their families?

Religious or spiritual practices can offer comfort, meaning, and routine to individuals with Alzheimer's and their families. Attending religious services, praying, or meditating can provide emotional support and a sense of connection. For many, spirituality helps reduce anxiety and provides a framework for coping with the challenges of the disease.

Q33: How can family dynamics be managed when caring for someone with Alzheimer's?

Family dynamics can be managed by fostering open communication, dividing caregiving responsibilities, and respecting each family member's strengths and limitations. It's also important to acknowledge differing opinions on care decisions and offer emotional support to one another throughout the caregiving process.

Q34: What financial support options are available for families affected by Alzheimer's disease?

Financial support options for Alzheimer's patients and their families include long-term care insurance, Medicaid, veterans' benefits, and local support programs. Consulting with a financial advisor specializing in elder care can also be helpful in planning for the costs associated with Alzheimer's treatment and care.

Q35: What are the legal rights of Alzheimer's patients in care settings?

Alzheimer's patients have the right to dignity, respect, and quality care, as outlined in laws such as the Americans with Disabilities Act (ADA). They are entitled to privacy, freedom from abuse or neglect, and informed consent in treatment decisions, with caregivers serving as advocates when necessary. Legal protections also extend to family members, ensuring they receive adequate care and support.

Q36: How can early planning benefit families caring for someone with Alzheimer's?

Early planning helps families understand their loved one's wishes, make necessary legal and financial arrangements, and develop a structured care plan. It reduces stress, ensures decisions are made in advance, and enables families to manage the disease's progression while focusing on providing the best care possible.

Q37: What are effective strategies for managing sleep disturbances in Alzheimer's patients?

Strategies for managing sleep disturbances include:

- Establishing a regular sleep routine.
- Reducing caffeine intake.
- Creating a calm, dark environment for bedtime.

Encouraging daytime physical activity and limiting naps can also improve sleep quality. In some cases, medications may be prescribed to address sleep issues, but they should be used cautiously.

Q38: What should caregivers consider when traveling with someone with Alzheimer's?

When traveling with someone who has Alzheimer's, caregivers should plan for a slower pace, ensure the destination is safe, and prepare for potential behavioral changes. It's essential to have necessary medications, medical information, and emergency contacts readily available. Creating a comfortable and familiar environment will help reduce stress and confusion for the patient.

Q39: How do caregivers adapt to the changing needs of someone with Alzheimer's over time?

Caregivers adapt by regularly reassessing the individual's abilities and adjusting care strategies accordingly. As the disease progresses, caregiving duties may become more physically demanding and emotionally challenging. It's important for caregivers to seek support, take breaks, and modify the home environment to ensure safety and comfort.

Q40: How can families plan for the transition to long-term care for a loved one with Alzheimer's?

Planning for long-term care involves assessing the patient's needs, exploring care facility options, and addressing financial and legal matters. Families should plan ahead to minimize stress during the transition and ensure the selected facility is suitable for the patient's stage of Alzheimer's. Effective communication and preparation are essential for a smooth transition.

Q41: What is the impact of seasonal changes in nature on the mood and behavior of individuals with Alzheimer's?

Seasonal changes in nature can significantly affect the mood and behavior of individuals with Alzheimer's disease. During spring and summer, longer days and increased sunlight help regulate sleep patterns and improve mood, while pleasant weather encourages outdoor activities like walking or gardening. These activities provide physical exercise and sensory stimulation, triggering positive memories and reducing confusion or distress. The vibrant natural environment during these seasons generally promotes well-being and comfort.

In contrast, the cooler temperatures and shorter days of fall and winter can limit outdoor activity, leading to feelings of isolation and sadness. Reduced sunlight in winter can disrupt circadian rhythms, worsen sleep patterns, and contribute to depression or cognitive decline. The lack of sensory stimulation during these months may increase agitation and anxiety. To mitigate these effects, caregivers can help individuals with Alzheimer's stay engaged with nature and maximize outdoor time throughout the year.

–

In conclusion, living with Alzheimer's disease requires a multifaceted approach that encompasses both the practical and emotional aspects of daily life. By implementing strategies to manage the disease's physical, cognitive, and behavioral challenges, individuals with Alzheimer's and their caregivers can create an environment that fosters independence, dignity, and emotional well-being. The importance of early planning, effective communication, and utilizing available resources cannot be overstated, as these factors help reduce stress, enhance quality of life, and ensure that both patients and caregivers feel supported throughout the journey. With the right tools, compassion, and a strong support network, individuals living with Alzheimer's disease can continue to experience moments of joy, connection, and purpose, while caregivers can find fulfillment in providing meaningful care and maintaining their own well-being. Through resilience, preparation, and community, the challenges of Alzheimer's can be met with strength and hope.

Bibliography

1. ***Anderson, M. S., et al. (2020).*** *"Behavioral interventions for Alzheimer's management using technology: Home-based monitoring." Current Geriatrics Reports, 9(2), 90–100, https://doi.org/10.1007/s13670-020-00312-y.*
2. ***Atri, A . (2019, April).*** *"Current and future treatments in Alzheimer's disease." Seminars in Neurology, 39(2), 227–240, doi:10.1055/s-0039-1678581.*
3. ***Brodaty, H., & Donkin. M. (2009).*** *Family caregivers of people with dementia. Dialogues Clin Neurosci., 11(2), 217–2 28.*
4. ***Gaugler, J. E., Kane, R. L., Langwell, K. M., & Newcomer, R. (2007).*** *The effects of dementia caregiving on work and retirement. Gerontologist , 47(6), 747 –7 59.*
5. ***Piers, R., et al. (2018, June 21).*** *" Advance care planning in dementia: Recommendations for healthcare professionals. " BMC Palliative Care, 17(1), 88 . Gale Academic OneFile, dx.doi.org.libaccess.lib.mcmaster.ca/10.1186/s12904-018-0332-2. Accessed 10 Nov. 2024.*
6. ***Phinney, A., & Moody, E. M. (2011).*** *"Leisure connections: Benefits and challenges of participating in a social recreation group for people with early dementia." Activities, Adaptation, & Aging, 35(2), 111–130, https://doi.org/10.1080/01924788.2011.572272.*
7. ***Robertson, K. J., et al. (2022).*** *"Barriers and facilitators of using dementia community support services provided by an Alzheimer's organisation: Perceptions of informal caregivers receiving services." Health & Social Care in the Community, 30(6), 2353–2361, https://doi.org/10.1111/hsc.13796.*
8. ***Zarit, S. H., Reever, K. E., & Bach-Peterson, J. (1980).*** *Relatives of the impaired elderly: Correlates of feelings of burden. Gerontologist , 20(6), 649– 655.*

Addressing Common Concerns and Misconceptions

CHAPTER 6

Addressing common concerns and misconceptions about Alzheimer's disease is essential for fostering a better understanding of this complex condition and reducing the stigma surrounding it. Misunderstandings about Alzheimer's often lead to unnecessary fear, delayed diagnosis, and barriers to effective care. This discussion begins by clarifying the difference between Alzheimer's and normal aging, debunking myths such as the belief that memory loss is an inevitable part of growing older. It also addresses misconceptions about risk factors, offering evidence-based insights into genetic, lifestyle, and environmental contributors to the disease.

Practical concerns, such as the effectiveness of treatments, the possibility of prevention, and the role of caregiving, are explored in detail to provide readers with a balanced perspective. The emotional aspects of Alzheimer's, including fears of losing independence or becoming a burden to loved ones, are also acknowledged, along with strategies for overcoming these challenges. By tackling these common concerns and correcting widespread misconceptions, this exploration empowers individuals and families to approach Alzheimer's with greater confidence, understanding, and compassion.

Q1: Is Alzheimer's disease a normal part of aging?

No, Alzheimer's disease is not a normal part of aging. While age is a significant risk factor for developing Alzheimer's, it is not an inevitable consequence of growing older. Most older adults do not develop Alzheimer's disease, and age-related cognitive changes are distinct from the progressive decline seen in Alzheimer's.

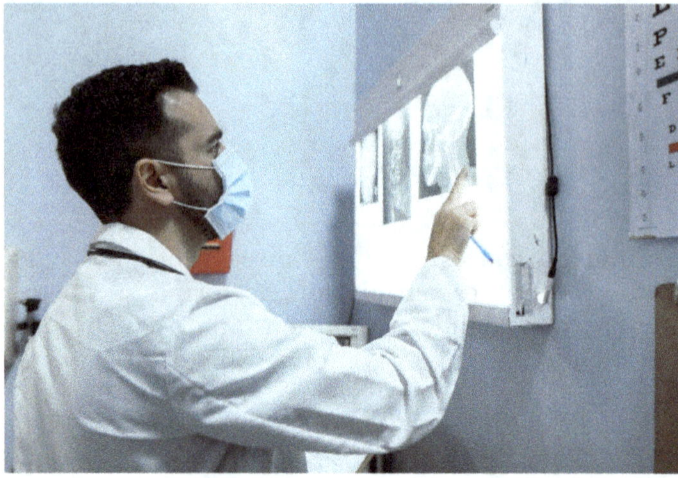

▲ **FIGURE 6.1.**
Doctor looking at an x-ray image.

Q2: Can Alzheimer's disease be prevented?

While there is no guaranteed way to prevent Alzheimer's disease, certain lifestyle factors may reduce the risk of developing the condition. These include maintaining a healthy diet, staying physically active, engaging in cognitive stimulation, managing cardiovascular risk factors, and staying socially connected. However, genetics and other factors also play a role, and not all cases of Alzheimer's can be prevented.

Q3: Do aluminum exposure and antiperspirants increase the risk of Alzheimer's disease?

There is no conclusive evidence linking aluminum exposure or the use of antiperspirants to an increased risk of Alzheimer's disease. While some studies have suggested a potential association between aluminum and Alzheimer's, the overall scientific consensus is that aluminum is unlikely to be a significant risk factor for the disease. More research is needed to fully understand the role, if any, of aluminum in Alzheimer's.

Q4: Are there effective dietary supplements for preventing or treating Alzheimer's disease?

There is currently no substantial evidence to support the use of dietary supplements for preventing or treating Alzheimer's disease. While certain nutrients, such as omega-3 fatty acids, antioxidants, and vitamins, are essential for brain health, supplementing with high doses has not been consistently shown to prevent or slow the progression of Alzheimer's. Focusing on a balanced diet and consulting with a healthcare provider before starting any supplements is essential.

Q5: Can exposure to electromagnetic fields (EMFs) from electronic devices increase the risk of Alzheimer's disease?

There is limited evidence to suggest that exposure to electromagnetic fields (EMFs) from electronic devices may affect brain health, but the link to Alzheimer's disease is uncertain. Some studies have reported associations between EMF exposure and neurological symptoms. Still, the evidence is inconclusive, and more research is needed to determine any potential long-term effects on brain function.

Q6: Is there a connection between Alzheimer's disease and the use of statin medications?

The relationship between statin medications, which lower cholesterol levels, and Alzheimer's disease is complex and not fully understood. While some studies have suggested a potential protective effect of statins against Alzheimer's, others have found no significant association or even an increased risk. Further research is needed to clarify the relationship between statins and Alzheimer's and to understand any underlying mechanisms.

Q7: Do traumatic brain injuries (TBIs) increase the risk of Alzheimer's disease?

There is growing evidence to suggest that traumatic brain injuries (TBIs), especially repeated or severe injuries, may increase the risk of developing Alzheimer's disease later in life. TBIs can lead to the accumulation of abnormal proteins in the brain, including amyloid plaques and tau tangles, which are characteristic of Alzheimer's pathology. Minimizing the risk of TBIs through prevention strategies, such as wearing helmets and practicing safe sports and activities, may help reduce the risk of Alzheimer's.

Q8: Can exposure to pesticides and other environmental toxins increase the risk of Alzheimer's disease?

Exposure to pesticides and other environmental toxins has been studied as a potential risk factor for Alzheimer's disease, but the evidence is inconclusive. Some studies have reported associations between pesticide exposure and an increased risk of Alzheimer's, while others have found no significant link. More research is needed to understand the role of environmental toxins in Alzheimer's and to develop strategies for minimizing exposure.

Q9: Is there a connection between sleep disturbances and Alzheimer's disease?

There is growing evidence to suggest that sleep disturbances, such as insomnia, sleep apnea, and circadian rhythm disruptions, may increase the risk of Alzheimer's disease and accelerate cognitive decline in individuals already affected by the condition. Sleep plays a crucial role in brain health, including the clearance of toxic proteins like amyloid beta, so optimizing sleep quality and duration may help reduce the risk of Alzheimer's and improve cognitive function.

Q10: Can chronic stress increase the risk of Alzheimer's disease?

Chronic stress has been implicated in Alzheimer's disease risk by promoting inflammation, oxidative stress, and changes in brain structure and function. Over time, prolonged exposure to stress hormones like cortisol may also impair memory and cognitive function. While more research is needed to fully understand the relationship between chronic stress and Alzheimer's, managing stress through relaxation techniques, social support, and healthy coping strategies may be beneficial for brain health.

Q11: Is there a genetic test for predicting Alzheimer's disease risk?

Genetic tests can assess an individual's risk of developing Alzheimer's disease, primarily by analyzing gene variations like APOE. However, these tests cannot definitively predict whether someone will develop the disease, as genetics is just one of many factors that influence Alzheimer's risk. Genetic testing for Alzheimer's disease risk should be done in consultation with a healthcare provider and with careful consideration of potential implications for the individual and their family.

Q12: Can mental and social activities reduce the risk of Alzheimer's disease?

Engaging in mentally stimulating activities, such as reading, puzzles, learning new skills, and maintaining social connections, has been associated with a reduced risk of Alzheimer's disease. Mental and social activities promote brain health by stimulating neural connections, enhancing cognitive reserve, and reducing stress and depression. Incorporating these activities into daily life may help support cognitive function and reduce the risk of Alzheimer's as we age.

Q13: Are there lifestyle factors that can increase the risk of Alzheimer's disease?

Several lifestyle factors have been associated with an increased risk of Alzheimer's disease, including a sedentary lifestyle, poor diet, smoking, excessive alcohol consumption, and untreated chronic health conditions like obesity, diabetes, and hypertension. These factors can contribute to cardiovascular risk factors, inflammation, and oxidative stress, which in turn may increase the risk of Alzheimer's. Adopting a healthy lifestyle that includes regular physical activity, a balanced diet, and management of chronic health conditions may help reduce the risk of Alzheimer's and promote brain health.

Q14: Does education level impact the risk of Alzheimer's disease?

Higher levels of education have been associated with a reduced risk of Alzheimer's disease and other forms of dementia. Education contributes to cognitive reserve, which refers to the brain's ability to withstand neuropathological damage and maintain mental function despite age-related changes or brain injury. Lifelong learning, intellectual stimulation, and cognitive engagement associated with higher education may help build cognitive reserve and delay the onset of Alzheimer's symptoms.

Q15: Is there a connection between depression and Alzheimer's disease?

Depression has been associated with an increased risk of Alzheimer's disease and may also exacerbate cognitive decline in individuals already affected by the condition. The relationship between depression and Alzheimer's is complex and bidirectional, with overlapping biological mechanisms and risk factors. Treating depression and addressing mental health concerns may help reduce the risk of Alzheimer's and improve overall well-being.

Q16: Is there a link between heart health and Alzheimer's disease?

Heart health and brain health are closely interconnected, and cardiovascular risk factors such as high blood pressure, high cholesterol, diabetes, and obesity have been associated with an increased risk of Alzheimer's disease. These risk factors can damage blood vessels, impair circulation, and contribute to inflammation and oxidative stress in the brain, potentially accelerating cognitive decline. Managing cardiovascular risk factors through lifestyle modifications and medical interventions may help reduce the risk of Alzheimer's and promote overall brain health.

Q17: Can hormonal changes, such as those during menopause or andropause, affect Alzheimer's disease risk?

Hormonal changes, such as those occurring during menopause in women and andropause (age-related decline in testosterone levels) in men, may influence Alzheimer's disease risk. Estrogen has neuroprotective effects and may help maintain cognitive function, so declining estrogen levels during menopause may increase the risk of Alzheimer's in women. Similarly, testosterone may play a role in brain health, and reductions in testosterone levels during andropause may affect cognitive function in men. However, the relationship between hormonal changes and Alzheimer's risk is complex and requires further research for a better understanding.

Q18: Can exposure to air pollution increase the risk of Alzheimer's disease?

Exposure to air pollution, especiallyparticulate matter (PM2.5) and other pollutants has been associated with an increased risk of Alzheimer's disease and other neurodegenerative disorders. Air pollution can contribute to inflammation, oxidative stress, and vascular damage in the brain, potentially accelerating the onset and progression of Alzheimer's pathology. Minimizing exposure to air pollution through environmental regulations, urban planning, and personal protective measures may help reduce the risk of Alzheimer's and promote brain health.

Q19: Are there ethnic or racial disparities in Alzheimer's disease risk?

Ethnic and racial disparities exist in Alzheimer's disease risk, with higher rates of the condition observed among specific population groups, including African Americans, Hispanics, and Native Americans. Socioeconomic factors, access to healthcare, educational attainment, genetic predisposition, and cultural differences in lifestyle and health behaviors may influence these disparities. Addressing health inequities and promoting culturally sensitive approaches to Alzheimer's prevention, diagnosis, and care are essential for reducing inequalities and improving outcomes for all individuals affected by the disease.

Q20: Can Alzheimer's disease be transmitted from person to person, like an infection?

Alzheimer's disease is not considered transmissible from person to person, like an infection. While abnormal proteins associated with Alzheimer's, such as amyloid beta and tau, can spread within the brain and contribute to disease progression, there is no evidence to suggest that it can be transmitted through casual contact or exposure to bodily fluids. The vast majority of Alzheimer's cases are sporadic, with complex interactions between genetic, environmental, and lifestyle factors contributing to disease development.

Q21: Is there a relationship between Alzheimer's disease and exposure to certain medications?

Some medications have been associated with an increased risk of cognitive impairment or dementia, especially in older adults. These medications include certain anticholinergic drugs (used to treat allergies, insomnia, and depression), benzodiazepines (used to treat anxiety and sleep disorders), and proton pump inhibitors (used to treat acid reflux). However, the relationship between medication uses and Alzheimer's disease risk is complex, and more research is needed to understand potential mechanisms and identify safer alternatives for vulnerable populations.

Q22: Can Alzheimer's disease be diagnosed through a simple blood test?

Currently, there is no simple blood test available for diagnosing Alzheimer's disease. While researchers actively investigate blood-based biomarkers for Alzheimer's, such as amyloid beta and tau protein levels, no single biomarker has been validated for routine clinical use. Diagnosing Alzheimer's typically involves a comprehensive evaluation, including medical history, cognitive assessments, neuroimaging scans, and sometimes cerebrospinal fluid analysis. Advances in blood-based biomarker research promise to improve early detection and diagnosis of Alzheimer's in the future.

Q23: Is Alzheimer's disease more common in women than in men?

Alzheimer's disease is more common in women than in men, with women accounting for nearly two-thirds of Alzheimer's cases worldwide. Some researchers attribute this gender disparity to differences in longevity, as women tend to live longer than men, and age is the most significant risk factor for Alzheimer's. However, hormonal factors, genetic predisposition, and social determinants of health may also contribute to the higher prevalence of Alzheimer's among women. Understanding these gender differences is essential for tailoring prevention and intervention strategies to address the needs of both men and women affected by Alzheimer's.

Q24: Can Alzheimer's disease be prevented or reversed through brain-training games and apps?

While brain-training games and apps have gained popularity to maintain cognitive function and prevent Alzheimer's disease, their effectiveness remains uncertain. While engaging in mentally stimulating activities may offer some mental benefits, especially when combined with other healthy lifestyle habits, there is limited evidence to support the claim that brain-training games alone can prevent or reverse Alzheimer's. More rigorous research is needed to evaluate the long-term effects of brain-training interventions on cognitive function and brain health.

Q25: Is Alzheimer's disease the same as dementia?

Alzheimer's disease is the most common cause of dementia, accounting for an estimated 60–80% of dementia cases. However, dementia is an umbrella term that encompasses a group of symptoms associated with a decline in cognitive function severe enough to interfere with daily life. Other common causes of dementia include vascular dementia, Lewy body dementia, and frontotemporal dementia. While specific neuropathological changes characterize Alzheimer's, the clinical presentation of dementia can vary depending on the underlying cause.

Q26: Can alcohol consumption increase the risk of Alzheimer's disease?

Moderate alcohol consumption has been associated with a lower risk of cognitive decline and dementia in some studies. In contrast, heavy alcohol consumption is a known risk factor for cognitive impairment and dementia. Excessive alcohol consumption can damage brain cells, impair mental function, and increase the risk of conditions like alcohol-related dementia and Wernicke-Korsakoff syndrome. While the relationship between alcohol and Alzheimer's disease risk is complex, moderation is key, and individuals should consult with a healthcare provider regarding alcohol consumption and brain health.

Q27: Can Alzheimer's disease be detected through changes in the sense of smell?

Changes in the sense of smell have been observed in individuals with Alzheimer's and other neurodegenerative disorders, but they are not specific enough to serve as a diagnostic marker. While olfactory dysfunction may occur in the early stages of Alzheimer's, it can also be caused by other factors such as aging, smoking, and certain medical conditions. Research into olfactory biomarkers for Alzheimer's is ongoing, but currently, diagnosis typically relies on a combination of clinical evaluations and specialized tests.

Q28: Can alternative therapies or supplements treat Alzheimer's disease?

While some alternative therapies and supplements have been proposed for treating Alzheimer's disease, such as herbal remedies, acupuncture, and dietary supplements, there is limited scientific evidence to support their efficacy in treating the underlying causes of Alzheimer's or slowing disease progression. While certain herbs and supplements may have antioxidant or anti-inflammatory properties that could theoretically benefit brain health, rigorous clinical trials have not consistently demonstrated significant effects on Alzheimer's symptoms or outcomes. It's essential for individuals considering alternative therapies to consult with a healthcare provider and approach these treatments with caution, as they may interact with medications or have unforeseen risks.

Q29: Can Alzheimer's disease be inherited?

While genetic factors can influence the risk of developing Alzheimer's disease, most cases are not directly inherited in a simple Mendelian fashion. Late-onset Alzheimer's disease, which typically occurs after age 65, has a complex genetic component involving multiple genes and environmental factors. The most potent genetic risk factor for late-onset Alzheimer's is the APOE gene, specifically the APOE4 allele. However, having this gene variant does not guarantee that an individual will develop the disease. Early-onset Alzheimer's, which occurs before age 65, is more strongly influenced by genetic factors, including mutations in genes such as APP, PSEN1, and PSEN2. Still, it accounts for only a tiny percentage of all Alzheimer's cases.

Q30: Is there a cure for Alzheimer's disease?

There is currently no cure for Alzheimer's disease. While some medications may temporarily alleviate symptoms or slow disease progression in some individuals, they do not address the underlying causes of Alzheimer's or stop the disease from advancing. Research into potential disease-modifying treatments and preventive strategies is ongoing, with promising avenues including immunotherapy, gene therapy, lifestyle interventions, and precision medicine approaches. However, developing effective treatments for Alzheimer's remains a significant scientific challenge, and more research is needed to achieve breakthroughs in diagnosis, treatment, and, ultimately, a cure for the disease.

Q31: Can Alzheimer's disease be treated with antibiotics?

Bacterial infections do not cause Alzheimer's disease, so antibiotics are not an appropriate treatment. While some studies have suggested a possible link between infections, such as chronic periodontitis or respiratory infections, and Alzheimer's risk, the evidence is inconclusive. Antibiotics should only be used to treat bacterial infections diagnosed by a healthcare professional and not as a preventive measure or treatment for Alzheimer's disease. Consulting with a healthcare provider is essential for the appropriate diagnosis and management of Alzheimer's and related conditions.

Q32: Are there clinical trials available for Alzheimer's disease treatment?

Yes, ongoing clinical trials are investigating potential treatments for Alzheimer's disease at various stages of development. These trials evaluate new drugs, interventions, and preventive strategies aimed at slowing or stopping the progression of the disease, improving symptoms, and enhancing the quality of life for individuals affected by Alzheimer's. Clinical trials may involve testing investigational drugs, lifestyle interventions, cognitive training programs, or combinations of treatments. Participating in clinical trials can provide access to cutting-edge therapies and contribute to advancing Alzheimer's research. Still, it's essential to carefully consider the potential risks and benefits before enrolling in a trial.

Q33: Can Alzheimer's disease be reversed with lifestyle changes?

While lifestyle changes such as regular physical activity, healthy diet, cognitive stimulation, and social engagement can promote brain health and potentially slow the progression of Alzheimer's disease, they cannot reverse the condition once it has developed. Alzheimer's is a progressive neurodegenerative disorder characterized by irreversible damage to brain cells and the accumulation of abnormal proteins. However, adopting a healthy lifestyle can help optimize cognitive function, improve quality of life, and potentially delay the onset of symptoms in individuals at risk for Alzheimer's.

Q34: Can Alzheimer's disease cause hallucinations or delusions?

Yes, Alzheimer's disease can cause hallucinations or delusions, especially in the later stages of the disease. Hallucinations involve seeing, hearing, or feeling things that are not present, while delusions are false beliefs or perceptions that are firmly held despite evidence. These symptoms can be distressing for both the individual with Alzheimer's and their caregivers, and they may result from changes in brain function, alterations in perception and reality processing, or a response to stress or unfamiliar environments. Managing hallucinations and delusions often requires a multifaceted approach, including environmental modifications, behavioral interventions, and sometimes medications to address underlying factors contributing to these symptoms.

Q35: Can Alzheimer's disease cause aggression or agitation?

Yes, Alzheimer's disease can cause aggression, agitation, or other behavioral changes, particularly as the disease progresses. These behaviors may result from confusion, frustration, discomfort, or an inability to communicate needs effectively. Changes in brain chemistry and structure, as well as alterations in mood-regulating neurotransmitters, may also contribute to behavioral symptoms in Alzheimer's. Managing aggression and agitation involves:

- Identifying triggers.
- Creating a calm and supportive environment.
- Implementing structured routines.
- Sometimes, medications or other interventions address underlying causes and alleviate distress.

Q36: Can Alzheimer's disease cause wandering or getting lost?

Yes, wandering and getting lost are common behavioral symptoms of Alzheimer's disease, especially in the middle and later stages of the condition. Individuals with Alzheimer's may wander, become disoriented, or attempt to leave familiar surroundings without a clear destination or awareness of their surroundings. Wandering can pose safety risks, including the potential for injury or loss, so caregivers need to implement strategies to prevent wandering and ensure the safety of their loved ones. This may include using door alarms, providing identification, and enrolling in programs like Medic Alert® + Alzheimer's Association Safe Return®.

Q37: Can Alzheimer's disease cause changes in eating habits or appetite?

Yes, Alzheimer's disease can cause changes in eating habits or appetite, leading to alterations in food preferences, meal patterns, and dietary intake. These changes may result from a variety of factors, including cognitive decline, sensory changes, difficulty with chewing or swallowing, medication side effects, or underlying medical conditions such as depression or gastrointestinal issues.

Maintaining proper nutrition and hydration is essential for overall health and well-being in individuals with Alzheimer's, so caregivers may need assistance with meal preparation, providing supportive dining environments, and monitoring nutritional intake to ensure adequate nourishment.

Q38: Can Alzheimer's disease cause sleep disturbances or insomnia?

Yes, Alzheimer's disease can cause sleep disturbances or insomnia, affecting both the quantity and quality of sleep in affected individuals. Sleep disturbances may manifest as difficulty falling asleep, frequent awakenings during the night, daytime sleepiness, or disturbances in the sleep– wake cycle. These sleep changes can result from alterations in circadian rhythms, brain function disruptions, melatonin production changes, or environmental factors. Managing sleep disturbances in Alzheimer's may involve:

- Establishing regular sleep routines.
- Creating a comfortable sleep environment.
- Promoting daytime activity and exposure to natural light.
- Addressing underlying medical or behavioral issues that may contribute to sleep problems.

Q39: Can Alzheimer's disease cause incontinence or bladder control problems?

Yes, Alzheimer's disease can cause incontinence or bladder control problems, leading to difficulties with urinary urgency, frequency, or leakage in affected individuals. These bladder changes may result from a combination of factors, including alterations in brain function affecting bladder control, decreased awareness of bodily sensations, mobility limitations, medication side effects, or underlying medical conditions such as urinary tract infections. Managing incontinence in Alzheimer's involves implementing toileting schedules, providing regular bathroom reminders, using absorbent products as needed, and addressing any underlying factors contributing to bladder dysfunction.

Q40: Can Alzheimer's disease cause difficulty with speech or communication?

Yes, Alzheimer's disease can cause difficulty with speech or communication, including problems with finding words, forming sentences, understanding language, or engaging in meaningful conversation. These language difficulties, known as aphasia, may result from changes in brain structure and function affecting areas responsible for language processing and declines in cognitive abilities such as memory, attention, and executive function. Supporting communication in Alzheimer's involves:

- Using simple language.
- Maintaining eye contact.
- Offering visual cues or gestures.
- Creating opportunities for social interaction and expression.

Q41: Can Alzheimer's disease cause falls or balance problems?

Yes, Alzheimer's disease can increase the risk of falls and balance problems, especially in the later stages of the condition. These mobility issues may result from changes in gait, muscle strength, coordination, and spatial awareness, as well as cognitive impairments affecting judgment and decision-making. Falls pose significant risks for individuals with Alzheimer's, including injuries such as fractures, head trauma, and complications from immobility. Preventing falls in Alzheimer's involves:

- Removing hazards from the environment.
- Providing mobility aids as needed.
- Implementing exercise programs to improve strength and balance.
- Supervising activities to ensure safety.

Q42: Can Alzheimer's disease cause apathy or lack of interest in activities?

Yes, Alzheimer's disease can cause apathy, characterized by a lack of interest, motivation, or initiative in activities that were previously enjoyable or meaningful to the individual. Apathy may result from changes in brain function affecting reward pathways, neurotransmitter imbalances, or emotional processing deficits in Alzheimer's. This loss of interest can significantly impact the quality of life for individuals with Alzheimer's, as well as their caregivers. Addressing apathy may involve providing structure and encouragement, simplifying tasks, offering meaningful activities, and addressing any underlying factors contributing to apathetic behavior.

Q43: Can Alzheimer's disease cause mood swings or emotional changes?

Yes, Alzheimer's disease can cause mood swings or emotional changes, leading to fluctuations in mood, behavior, and affect in affected individuals. These emotional changes may manifest as irritability, agitation, anxiety, depression, or emotional lability (rapid shifts in emotions). Changes in brain chemistry, alterations in neurotransmitter levels, and disruptions in neural circuits regulating emotion may contribute to mood swings in Alzheimer's. Managing emotional changes may involve:

- Addressing underlying factors such as pain or discomfort.
- Providing emotional support and reassurance.
- Using behavioral interventions or medications as needed to stabilize mood and improve well-being.

Q44: Can Alzheimer's disease cause difficulties with swallowing or eating?

Yes, Alzheimer's disease can cause difficulties with swallowing or eating, known as dysphagia, particularly in the later stages of the condition. Dysphagia may result from changes in oral motor function, coordination of swallowing muscles, or sensory processing deficits affecting the ability to chew, swallow, and safely manage food and liquids. Dysphagia poses risks for choking, aspiration pneumonia, malnutrition, and dehydration in individuals with Alzheimer's. Managing dysphagia involves:

- Modifying food and liquid textures.
- Providing swallowing therapy.
- Using adaptive feeding utensils.
- Ensuring proper positioning during meals to minimize risks and promote safe eating and swallowing.

Q45: Can Alzheimer's disease cause changes in personality or behavior?

Yes, Alzheimer's disease can cause changes in personality or behavior, leading to alterations in mood, social interactions, and interpersonal relationships in affected individuals. These changes may manifest as irritability, agitation, apathy, disinhibition, impulsivity, or socially inappropriate behaviors. Changes in brain structure and function and disruptions in neural circuits regulating emotion and behavior may contribute to personality changes in Alzheimer's. Managing behavioral changes in Alzheimer's involves:

- Identifying triggers.
- Providing a structured and supportive environment.
- Offering reassurance and validation.
- Using behavioral interventions to address specific symptoms.

In some cases, medications may be prescribed to manage behavioral symptoms and improve the quality of life for individuals with Alzheimer's.

Q46: Can Alzheimer's disease cause visual or spatial perception difficulties?

Yes, Alzheimer's disease can cause visual or spatial perception difficulties, leading to problems with interpreting visual information, judging distances, or perceiving objects in the environment accurately. These perceptual changes may result from alterations in brain regions responsible for visual processing and spatial awareness and declines in cognitive abilities such as attention and executive function. Visual and spatial perception difficulties can impact mobility, navigation, and safety for individuals with Alzheimer's, so creating clear and clutter-free environments, providing cues or markers for orientation, and addressing any underlying visual impairments or cognitive deficits contributing to perceptual difficulties are essential.

Bibliography

1. ***Carrillo, M. C., Brashear, H. R., Logovinsky, V., Ryan, J. M., Feldman, H. H., Siemers, E. R., et al. (2013).*** *Can we prevent Alzheimer 's disease? Secondary "prevention" trials in Alzheimer's disease. J Alzheimer's Dis., 33(s1).*
2. ***Fargo, K., & Bleiler, L. (2014).*** *Alzheimer's Association Report: 2014 Alzheimer's disease facts and figures. Alzheimer's Dement. 10(2), e47–e92.*
3. ***Mahoney, R., Regan, C., Katona, C., & Livingston, G. (2005).*** *Anxiety and depression in family caregivers of people with Alzheimer's disease: The LASER-AD study. American Journal of Geriatric Psychiatry , 13(9), 795– 801.*

Alzheimer's Disease Caregiving

CHAPTER 7

Alzheimer's disease caregiving is a journey that requires patience, adaptability, and a deep well of compassion. Caregivers play a critical role in supporting individuals with Alzheimer's as they face the physical, cognitive, and emotional challenges of the disease. This discussion begins by exploring the evolving responsibilities of caregiving, from assisting with daily tasks and managing medical needs to navigating complex behavioral changes such as agitation, confusion, or aggression. Practical strategies are provided to ensure a safe and supportive environment, including tips for effective communication, managing routines, and preventing accidents.

The emotional toll of caregiving is also addressed, with a focus on recognizing signs of caregiver burnout and accessing support systems such as respite care, counseling, and local resources. Legal and financial considerations, including power of attorney and long-term care planning, are highlighted to help caregivers prepare for future stages of the disease. Additionally, this discussion emphasizes the importance of self-care, encouraging caregivers to prioritize their own mental and physical health. By offering guidance on these essential aspects, this exploration equips caregivers with the tools and resources needed to provide compassionate, effective care while maintaining their own well-being.

Q1: What is Alzheimer's disease caregiving?

Alzheimer's disease caregiving involves providing physical, emotional, and practical support to individuals living with Alzheimer's disease or other forms of dementia. Caregivers assist with activities of daily living, manage medications, coordinate medical appointments, provide companionship, and ensure the safety and well-being of their loved ones.

▲ **FIGURE 7.1.**
Sharing moments together.

Q2: Who are Alzheimer's disease caregivers?

Alzheimer's disease caregivers can be family members, spouses, adult children, friends, or professional caregivers who assume responsibility for supporting and caring for individuals with Alzheimer's disease. Depending on the needs and preferences of the care recipient, caregivers may assist on a part-time or full-time basis.

▲ **FIGURE 7.2.**
Son cares for elderly father with Alzheimer's disease.

Q3: What are the challenges of Alzheimer's disease caregiving?

Alzheimer's disease caregiving presents various challenges, including emotional stress, physical exhaustion, financial strain, social isolation, role strain, and balancing caregiving responsibilities with other life demands such as work and family obligations. Caregivers may also experience grief, guilt, frustration, and anticipatory mourning as they witness the progression of the disease in their loved ones.

Q4: How does Alzheimer's disease caregiving impact caregivers' mental and physical health?

Alzheimer's disease caregiving can have significant impacts on caregivers' mental and physical health. Caregivers may experience symptoms of depression, anxiety, insomnia, chronic stress, fatigue, headaches, and other physical ailments due to the demands of caregiving and the emotional toll of witnessing their loved one's decline.

Q5: What are the strategies for managing stress and burnout in Alzheimer's disease caregiving?

Strategies for managing stress and burnout in Alzheimer's disease caregiving include:

- Practicing self-care.
- Seeking social support.
- Setting boundaries.
- Accepting help from others.
- Prioritizing tasks.
- Maintaining realistic expectations.
- Seeking respite care.
- Accessing counseling or support groups for caregivers.

Q6: How can caregivers promote independence and autonomy for individuals with Alzheimer's disease?

Caregivers can promote independence and autonomy for individuals with Alzheimer's disease by encouraging self-care activities, respecting their preferences and choices, involving them in decision-making, providing opportunities for meaningful engagement and social interaction, and adapting the environment to support their abilities and limitations.

Q7: What are the communication strategies for caregivers of individuals with Alzheimer's disease?

Communication strategies for caregivers of individuals with Alzheimer's disease include:

- Using simple language.
- Speaking slowly and clearly.
- Maintaining eye contact.
- Using nonverbal cues and gestures.
- Asking open-ended questions.
- Validating feelings and emotions.
- Practicing active listening to enhance understanding and connection.

Q8: How can caregivers address behavioral symptoms such as agitation and aggression in individuals with Alzheimer's disease?

Caregivers can address behavioral symptoms such as agitation and aggression in individuals with Alzheimer's disease by identifying triggers, creating a calm and structured environment, providing reassurance and validation, using distraction techniques, offering meaningful activities, and seeking professional help or medication management when necessary.

Q9: What are the safety considerations for caregivers of individuals with Alzheimer's disease?

Safety considerations for caregivers of individuals with Alzheimer's disease include:

- Preventing falls and accidents by removing hazards from the environment.
- Using assistive devices and mobility aids.
- Implementing home modifications.
- Ensuring proper medication management.
- Addressing wandering behavior through supervision and environmental changes.

Q10: How can caregivers manage sleep disturbances and sundowning in individuals with Alzheimer's?

Caregivers can manage sleep disturbances and sundowning in individuals with Alzheimer's disease by establishing a regular sleep routine, creating a comfortable sleep environment, promoting relaxation techniques, limiting caffeine and alcohol intake, providing daytime exposure to natural light, and consulting healthcare professionals for sleep assessment and management strategies.

Q11: What are the legal and financial considerations for Alzheimer's disease caregivers?

Legal and financial considerations for Alzheimer's disease caregivers include creating advance directives, power of attorney documents, and guardianship arrangements, managing financial affairs and healthcare decisions, accessing government assistance programs and benefits, and planning for long-term care needs and estate planning.

Q12: How can caregivers cope with the grief and loss associated with Alzheimer's disease caregiving?

Caregivers can cope with the grief and loss associated with Alzheimer's disease caregiving by acknowledging their feelings, seeking support from family and friends, participating in grief counseling or support groups, practicing self-compassion and self-care, finding meaning and purpose in caregiving experiences, and honoring the memory of their loved ones.

Q13: What are the resources available to Alzheimer's disease caregivers?

Resources available to Alzheimer's disease caregivers include caregiver support groups, helplines, online forums and communities, educational workshops and training programs, respite care services, adult day programs, memory cafes, and caregiver resource centers offered by local Alzheimer's associations and community organizations.

Q14: How can caregivers maintain their social connections and support networks while caregiving for individuals with Alzheimer's disease?

Caregivers can maintain their social connections and support networks while caregiving for individuals with Alzheimer's disease by staying connected with family and friends, scheduling regular outings and activities, joining social or recreational groups, attending caregiver support groups or educational events, and utilizing respite care services to take breaks and recharge.

Q15: What are the considerations for balancing work and caregiving responsibilities for individuals employed while caregiving for someone with Alzheimer's?

Balancing work and caregiving responsibilities for individuals employed while caring for someone with Alzheimer's disease requires careful planning, communication with employers, flexibility in work schedules, access to employee assistance programs and benefits, utilization of caregiving leave policies, and exploration of alternative work arrangements such as telecommuting or flexible hours.

Q16: How can caregivers address their loved one' s nutritional needs and dietary challenges in Alzheimer's disease?

Caregivers can address their loved one's nutritional needs and dietary challenges in Alzheimer's disease by offering nutrient-dense foods, providing small and frequent meals, encouraging hydration, accommodating nutritional preferences and restrictions, ensuring proper chewing and swallowing techniques, and consulting with healthcare professionals or dietitians for nutritional guidance.

Q17: What are the considerations for end-of-life care and advance care planning for individuals with Alzheimer's disease?

End-of-life care and advance care planning considerations for individuals with Alzheimer's disease include:

- Discussing treatment preferences and goals of care.
- Completing advance directives and living wills.
- Appointing a healthcare proxy or power of attorney.
- Establishing palliative care or hospice services.
- Addressing spiritual and emotional needs at the end of life.

Q18: How can caregivers support individuals with Alzheimer's disease in maintaining social connections and engagement?

Caregivers can help individuals with Alzheimer's disease maintain social connections and engagement by arranging visits with family and friends, participating in social outings and community events, volunteering or joining clubs and organizations, attending religious or cultural activities, and utilizing technology for virtual communication and social interaction.

Q19: What are the considerations for traveling with individuals with Alzheimer's disease?

Traveling with individuals with Alzheimer's disease requires careful planning, including preparing necessary documents and medications, choosing familiar and accessible destinations, scheduling rest breaks and accommodations, ensuring safety and comfort during transit, and anticipating and addressing potential challenges or triggers during the trip.

Q20: How can caregivers promote mental stimulation and cognitive engagement for individuals with Alzheimer's disease?

Caregivers can promote mental stimulation and cognitive engagement for individuals with Alzheimer's disease by providing activities that match their interests and abilities, such as puzzles, games, crafts, music, reading, and reminiscence activities. They can also encourage social interaction, storytelling, and participation in group activities to stimulate memory, communication, and emotional connections.

Q21: What are the considerations for managing medications and medical appointments for individuals with Alzheimer's disease?

Managing medications and medical appointments for individuals with Alzheimer's disease requires organization, communication with healthcare providers, and adherence to prescribed treatment plans. Caregivers should maintain a medication schedule, monitor for side effects or changes in symptoms, and communicate any concerns or observations to healthcare professionals.

Caregivers can address changes in behavior and personality in individuals with Alzheimer's disease by identifying triggers, maintaining a calm and supportive environment, validating feelings and emotions, offering reassurance and redirection, and utilizing behavioral interventions or medications recommended by healthcare professionals.

Q22: What are the considerations for providing personal care and hygiene assistance for individuals with Alzheimer's disease?

Providing personal care and hygiene assistance for individuals with Alzheimer's disease requires sensitivity, patience, and respect for their dignity and autonomy. Caregivers should establish routines, provide clear and straightforward instructions, help with grooming, bathing, dressing, and toileting as needed, and ensure privacy and comfort during personal care activities.

Q23: What are the considerations for culturally sensitive care for individuals with Alzheimer's and their families?

Providing culturally sensitive care for individuals with Alzheimer's disease and their families involves recognizing and respecting cultural beliefs, values, traditions, and preferences related to caregiving, end-of-life care, and treatment decisions. Caregivers should communicate openly, seek input from family members, and collaborate with cultural advisors or community leaders to provide culturally appropriate support and services.

Q24: How can caregivers prioritize their well-being and self-care while caring for individuals with Alzheimer's disease?

Caregivers can prioritize their well-being and self-care while caring for individuals with Alzheimer's disease by setting boundaries, asking for help, taking breaks and respite, maintaining social connections and hobbies, seeking counseling or support groups, practicing stress-reduction techniques, and acknowledging their needs and limitations.

Bibliography

1. ***Brodaty, H., & Donkin, M. (2009).*** *Family caregivers of people with dementia. Dialogues in Clinical Neuroscience, 11(2), 217–228.*
2. ***Schulz, R., & Martire, L. M. (2004).*** *Family caregiving of persons with dementia: Prevalence, health effects, and support strategies. American Journal of Geriatric Psychiatry, 12(3), 240–249.*
3. ***Pinquart, M., & Sörensen, S. (2003).*** *Differences between caregivers and noncaregivers in psychological health and physical health: A meta-analysis. Psychology and Aging, 18(2), 250–267.*

Social Support for Individuals with Alzheimer's

CHAPTER 8

In the journey of Alzheimer's disease, social support is an essential element for both individuals affected by the condition and their caregivers. Social connections and community resources play a crucial role in promoting well-being, managing symptoms, and enhancing the quality of life. As Alzheimer's progresses, the need for emotional, practical, and community-based support grows, requiring a collaborative effort among family, friends, healthcare providers, and social organizations.

This chapter explores the types of social support available for individuals with Alzheimer's, including caregiver support groups, respite services, and community programs. Through a question-and-answer format, we delve into strategies to reduce caregiver stress, maintain dignity and independence for individuals with Alzheimer's, and foster positive social interactions. By understanding these support mechanisms, families can navigate Alzheimer's with greater resilience and a sense of shared purpose.

Q1: How can caregivers cope with the stress and challenges of caring for someone with Alzheimer's disease?

Caregiving for someone with Alzheimer's disease can be emotionally and physically demanding. To cope with stress and

▲ FIGURE 8.1.
An elderly couple seeing a counselor.

challenges, caregivers should prioritize self-care, seek support from family and friends, join caregiver support groups, learn effective communication techniques, set realistic expectations, and consider respite care options to take breaks and recharge.

Q2: What resources are available to help caregivers of individuals with Alzheimer's disease?

Caregivers of individuals with Alzheimer's disease can access a variety of resources, including informational websites, educational materials, helplines, caregiver support groups, respite care services, adult day programs, and memory care facilities. Local Alzheimer's associations and community organizations often offer support services, educational workshops, and caregiver training programs tailored to the needs of caregivers.

▲ **FIGURE 8.2.**
Elderly woman searches for resources on a laptop

Q3: How can caregivers promote a sense of dignity and independence for individuals with Alzheimer's disease?

Caregivers can promote dignity and independence for individuals with Alzheimer's disease by involving them in decision-making, respecting their preferences and autonomy, maintaining familiar routines and surroundings, providing opportunities for meaningful activities and social engagement, and offering assistance and support in a sensitive and respectful manner.

Q4: What strategies can caregivers use to manage challenging behaviors in individuals with Alzheimer's disease?

Caregivers can manage challenging behaviors in individuals with Alzheimer's disease by identifying triggers, addressing unmet needs or discomfort, providing a calm and structured environment, using redirection and distraction techniques, validating feelings and emotions, and seeking professional help or behavioral interventions when necessary.

Q5: How can caregivers cope with the stress and challenges of caring for someone with Alzheimer's disease?

Caregivers can cope with the stress and challenges of caring for someone with Alzheimer's disease by prioritizing self-care practices such as regular exercise, adequate sleep, and healthy nutrition. Seeking support from family, friends, or caregiver support groups provides emotional validation and practical advice. Learning effective communication techniques helps in managing daily interactions and reducing frustration. Setting realistic expectations and accepting help from others can alleviate caregiver burden. Utilizing respite care options to take breaks and recharge is also crucial for maintaining physical and emotional well-being amidst the demands of caregiving.

Q6: What resources are available to help caregivers of individuals with Alzheimer's disease?

Caregivers of individuals with Alzheimer's disease have access to a range of valuable resources to support them in their caregiving roles. Informational websites like the Alzheimer's Association provide comprehensive guidance on understanding the disease and managing its challenges. Educational materials, including brochures and online publications, offer practical advice and caregiving tips. Helplines staffed by trained professionals offer confidential support and information about local resources. Caregiver support groups provide emotional validation and practical strategies through peer connections. Respite care services offer caregivers temporary relief, while adult day programs and specialized memory care facilities provide structured activities and specialized care for individuals with Alzheimer's. Local Alzheimer's associations often offer tailored support services, educational workshops, and community resources to meet caregivers' specific needs. These resources collectively aim to alleviate caregiver stress, enhance caregiving skills, and improve overall quality of life for both caregivers and individuals with Alzheimer's disease.

Q7: How can caregivers promote a sense of dignity and independence for individuals with Alzheimer's disease?

Caregivers can promote a sense of dignity and independence for individuals with Alzheimer's disease by involving them in decision-making processes, respecting their autonomy as much as possible, and maintaining familiar routines and environments. Offering opportunities for meaningful activities that align with their interests and abilities allows them to engage actively and maintain a sense of purpose. Caregivers should provide support sensitively and respectfully, allowing individuals to perform tasks independently when feasible and offering assistance with patience and understanding when needed. These approaches help individuals with Alzheimer's maintain their dignity and self-esteem while navigating daily challenges, enhancing their overall well-being and quality of life.

Q8: What strategies can caregivers use to manage challenging behaviors in individuals with Alzheimer's disease?

Caregivers can employ several strategies to manage challenging behaviors in individuals with Alzheimer's disease. Identifying triggers that lead to behaviors like agitation or confusion helps caregivers anticipate and mitigate these situations. Addressing the person's underlying needs promptly, such as hunger or discomfort, can prevent escalation. Creating a calm environment with minimal distractions and maintaining consistent routines provides stability and reduces anxiety. Redirecting attention to alternative activities or topics can help shift focus away from challenging behaviors, while validating the individual's emotions promotes understanding and reassurance. Caregivers may also benefit from seeking advice from healthcare professionals for specialized strategies or interventions. These approaches, applied with patience and empathy, can help caregivers effectively support individuals with Alzheimer's and manage difficult behaviors to enhance overall quality of care.

Q9: What role do caregiver support groups play in providing emotional and practical support?

Caregiver support groups play a crucial role in providing both emotional and practical support to individuals caring for someone with Alzheimer's disease. Emotionally, these groups offer a safe space for caregivers to express their feelings, share experiences, and receive empathy from others who understand the challenges they face. This validation of their emotions can reduce feelings of isolation and stress. Practically, support groups provide valuable information, tips, and strategies for managing caregiving responsibilities effectively. Caregivers can learn from each other's experiences, gain insights into navigating the complexities of Alzheimer's care, and access resources such as local services or caregiver workshops. Overall, caregiver support groups foster a sense of community, mutual support, and empowerment among caregivers, enhancing their ability to provide compassionate care while maintaining their own well-being.

Q10: How can caregivers balance their own needs with the demands of caregiving for someone with Alzheimer's disease?

Caregivers can balance their own needs with the demands of caregiving for someone with Alzheimer's disease by prioritizing self-care practices. This includes setting aside time for activities that promote physical and emotional well-being, such as exercise, adequate rest, and maintaining a healthy diet. Seeking support from family, friends, or caregiver support groups allows caregivers to share responsibilities and receive practical assistance. Effective time management and organization can help caregivers accomplish tasks while also allocating time for personal interests and relaxation. Utilizing respite care services periodically gives caregivers a break to recharge and attend to their own needs. It's essential for caregivers to communicate openly with their support network about their challenges and limitations, enabling others to offer assistance and relieve some of the caregiving burden. Balancing these aspects ensures caregivers can provide effective care while safeguarding their own health and well-being.

Q11: What are the benefits of respite care for caregivers of individuals with Alzheimer's disease?

Respite care offers several benefits for caregivers of individuals with Alzheimer's disease. Firstly, it provides caregivers with much-needed breaks from their caregiving responsibilities, allowing them time to rest, recharge, and attend to their own physical and emotional needs. This temporary relief helps prevent caregiver burnout and reduces stress, which in turn enhances their ability to provide quality care. Respite care also promotes social engagement and personal time, enabling caregivers to maintain relationships and participate in activities they enjoy. Importantly, it ensures the safety and well-being of the individual with Alzheimer's during the caregiver's absence, often in a professional care setting equipped to meet their specific needs. Overall, respite care plays a vital role in supporting caregivers' overall health, resilience, and capacity to sustain long-term caregiving.

Q12: How can caregivers encourage social interactions and meaningful activities for individuals with Alzheimer's disease?

Caregivers can encourage social interactions and meaningful activities for individuals with Alzheimer's disease by creating regular opportunities for social engagement with family and friends in familiar and comfortable settings. Group activities such as music therapy or art classes tailored to their abilities can stimulate cognitive function and provide enjoyable social interactions. Using memory aids and reminiscence therapy helps spark conversations and connect with past experiences. Virtual interactions through technology, like video calls or social apps, offer accessible ways to stay connected with loved ones. By adapting activities to the individual's interests and capabilities, caregivers can promote a sense of accomplishment and maintain their dignity while fostering positive social experiences. These efforts not only enhance quality of life but also help individuals with Alzheimer's feel valued and engaged in their communities.

Q13: What are the emotional and psychological effects of caregiving for someone with Alzheimer's disease?

Caregiving for someone with Alzheimer's disease can have significant emotional and psychological effects on caregivers.

Emotionally, caregivers often experience stress, anxiety, and sadness stemming from the challenges of witnessing their loved one's decline in cognitive abilities and functional independence. They may feel grief over the loss of the person they once knew and frustration with the unpredictable nature of the disease's progression. Caregivers may also experience guilt or resentment about their own feelings or limitations in providing care. Psychologically, caregiving can lead to heightened levels of caregiver burden, which includes feelings of overwhelm, exhaustion, and isolation. Caregivers may neglect their own needs, leading to physical and emotional strain, sleep disturbances, and changes in mood. The constant demands of caregiving can affect their ability to maintain relationships, pursue personal interests, or engage in social activities, further contributing to feelings of loneliness or depression. Despite these challenges, caregiving can also bring a sense of fulfillment and purpose as caregivers provide essential support and companionship to their loved ones. However, it's crucial for caregivers to recognize and address their emotional and psychological needs, seek support from others, and practice self-care to mitigate the negative effects of caregiving and maintain their own well-being.

Q14: How can caregivers effectively communicate with individuals who have Alzheimer's disease?

Caregivers can effectively communicate with individuals who have Alzheimer's disease by adopting strategies that accommodate the person's cognitive abilities and enhance understanding. Maintaining a calm and positive demeanor helps create a supportive environment, while using simple and clear language avoids confusion. Offering choices and listening actively to both verbal and nonverbal cues encourages engagement and empowers decision-making. Visual aids and gestures can supplement verbal communication, aiding comprehension. It's important to avoid arguments or corrections, instead validating emotions and redirecting conversations positively. Maintaining eye contact and creating a quiet environment minimizes distractions, facilitating focused communication. Above all, patience, flexibility, and empathy are essential in fostering meaningful interactions and supporting the individual's communication needs throughout the progression of Alzheimer's disease.

Q15: What are the signs of caregiver burnout and how can it be prevented?

Caregiver burnout, characterized by physical, emotional, and mental exhaustion, is common among those caring for individuals with Alzheimer's disease. Signs include constant fatigue, withdrawal from activities, irritability, and difficulty sleeping. To prevent burnout, caregivers should prioritize self-care by seeking support from family, friends, or support groups to share responsibilities and provide emotional relief. Setting realistic expectations, taking regular breaks, and practicing stress-relief techniques such as exercise or mindfulness are crucial. Delegating tasks and maintaining a balanced routine that includes personal time for hobbies and relaxation can also help prevent burnout. Monitoring one's own mental and physical health, recognizing signs of burnout early, and seeking professional assistance when needed are essential steps in maintaining resilience and providing effective care over the long term.

Q16: How do cultural factors influence caregiving practices and support for individuals with Alzheimer's disease?

Cultural factors profoundly shape caregiving practices and support for individuals with Alzheimer's disease. Cultural beliefs regarding family roles, respect for elders, and views on illness influence how caregiving responsibilities are distributed within families and communities. These beliefs often impact decisions about home versus institutional care, attitudes toward seeking professional assistance, and approaches to managing symptoms. Cultural norms also influence communication styles and preferences for spiritual or traditional practices in caregiving. Healthcare providers and support services can enhance care by understanding and respecting these cultural dynamics, ensuring that caregiving approaches are sensitive and aligned with the values and preferences of both caregivers and individuals with Alzheimer's disease.

Q17: What financial and legal considerations should caregivers be aware of when caring for someone with Alzheimer's disease?

Caregivers of individuals with Alzheimer's disease should address several key financial and legal considerations. Financial planning involves assessing and preparing for caregiving costs, accessing government benefits, and considering long-term care options. Legal preparations include establishing durable power of attorney and healthcare directives to manage decisions in case of incapacity. Estate planning ensures assets are organized, wills are updated, and financial affairs are managed effectively. Caregivers should also be vigilant against financial exploitation and seek professional advice from experts in elder law and financial planning to navigate these complexities and safeguard the well-being of both the individual with Alzheimer's and their family.

Q18: How can technology assist caregivers in managing the care of individuals with Alzheimer's disease?

Technology plays a crucial role in assisting caregivers of individuals with Alzheimer's disease by providing practical tools and resources. GPS tracking devices and smart home sensors help ensure safety by monitoring movements and alerting caregivers to potential risks like wandering. Medication management apps and electronic pill dispensers aid in administering medications on schedule, reducing errors. Communication tools such as video calling platforms and social media facilitate staying connected with healthcare providers and family members. Memory aids like reminder apps and digital calendars assist in daily task management and scheduling. Additionally, telehealth services enable caregivers to access medical support remotely, while online support groups and educational resources offer valuable guidance and emotional support. These technologies not only enhance caregiving efficiency but also improve the overall quality of life for both caregivers and individuals with Alzheimer's.

Q19: What community resources are available to support individuals with Alzheimer's disease and their caregivers?

Community resources for individuals with Alzheimer's disease and their caregivers offer essential support and assistance. Alzheimer's associations provide information, education, and caregiver training, while local support groups offer emotional support and practical advice. Memory care programs and respite care services provide supervised activities for individuals with Alzheimer's and temporary relief for caregivers. In-home care services and transportation assistance help maintain independence and access necessary medical appointments. Legal and financial counseling services offer guidance on planning and managing affairs, ensuring caregivers have the necessary tools to navigate the complexities of caregiving. These community resources collectively contribute to improving the quality of life for both caregivers and individuals living with Alzheimer's disease.

Q20: What are some communication tools or techniques caregivers can use to connect with individuals in advanced stages of Alzheimer's?

In the advanced stages of Alzheimer's, verbal communication becomes challenging, so caregivers can benefit from nonverbal techniques to maintain connection. Gentle touch, such as holding hands or giving a light shoulder pat, maintaining eye contact, and using expressive facial gestures can convey comfort and reassurance without words. Music therapy, particularly through familiar songs, has been shown to stimulate memory and evoke emotional responses, even in the late stages of the disease. Picture-based communication tools or memory books filled with photos of family, friends, and significant life events can prompt engagement and reminiscence. Simplifying language, speaking in a calm tone, and giving ample time for response can help the person feel less pressured. If verbal communication becomes difficult, looking for signs such as body movement or changes in facial expression can help you better understand their needs and emotional state. Using these approaches with patience and empathy helps bridge communication gaps, fostering a meaningful emotional connection.

Q21: How can caregivers prepare for the transition to memory care or long-term care facilities?

Preparing for the transition to memory care or long-term care facilities requires thoughtful planning to ease the process for both caregivers and individuals with Alzheimer's. Researching options early and visiting several facilities can help caregivers assess the quality of care, staff interaction, safety, and services available. Consulting with healthcare professionals, such as social workers or geriatric care managers, can provide valuable guidance and help narrow down suitable options. Engaging the individual with Alzheimer's in discussions about the transition early, if possible, allows caregivers to honor their preferences. Bringing familiar personal items, like photographs or favorite blankets, can help make the new environment feel more comfortable and familiar. Caregivers should also prepare emotionally, recognizing that adjusting to memory care may take time, and support from family members and counselors can help them through this transition.

Q23: How can caregivers address dietary and nutritional needs for individuals with Alzheimer's?

Meeting the dietary and nutritional needs of individuals with Alzheimer's can be challenging, particularly as they may experience appetite changes, forget to eat, or face difficulty swallowing. Providing small, frequent meals and nutrient-dense snacks throughout the day ensures that nutritional needs are met without overwhelming the person. Foods that are easy to eat, like finger foods and soft-textured options, help promote independence while minimizing frustration. Creating a calm, distraction-free environment during meals helps the person focus on eating and may reduce mealtime stress. Sharing meals together encourages social interaction and models healthy eating habits, fostering a more enjoyable experience. Proper hydration is essential, so offering water and other fluids regularly is vital. Consulting with a dietitian to tailor a balanced diet to the individual's specific needs is highly recommended.

Q24: How can caregivers help individuals with Alzheimer's maintain physical health and mobility?

Maintaining physical health and mobility is crucial for slowing cognitive decline and preserving a sense of independence in individuals with Alzheimer's. Caregivers can incorporate physical activities like walking, stretching, or simple balance exercises into daily routines, which not only improve physical health but can also uplift mood and reduce anxiety. Gentle activities like dancing, gardening, or seated exercises allow for safe movement and enjoyment without overexertion. Ensuring safety through environmental adjustments, such as installing handrails, non slip mats, and providing supportive footwear, helps prevent falls and encourages confident movement. Consulting a physical therapist can provide specific exercises to enhance mobility, improve balance, and reduce fall risks. These efforts contribute to physical health and foster a sense of independence and engagement in daily life.

Q25: How can caregivers support their children or young family members affected by a relative's Alzheimer's disease?

Caregivers can help children and young family members process the changes they see in a relative with Alzheimer's by providing clear, age-appropriate information about the disease. Explaining that Alzheimer's affects memory and behavior helps younger family members understand why their loved one might act differently. Open discussions about their feelings, fears, and questions are essential, allowing young family members to express emotions in a safe space. Encouraging positive interactions, such as playing simple games, working on arts and crafts, or viewing family photos, can help create joyful memories and strengthen connections. Support groups for families or individual counseling may offer additional guidance, helping younger individuals manage their emotions and build empathy, patience, and resilience in the face of these family changes.

Q26: What are some strategies for managing the end-of-life phase for individuals with Alzheimer's?

During the end-of-life phase, the focus shifts to providing comfort and ensuring a peaceful environment for individuals with Alzheimer's. Caregivers should prioritize physical comfort, managing any pain, and ensuring adequate hydration and comfort in positioning. Offering emotional support through gentle presence, soothing music, or reading aloud provides companionship and reassurance. Hospice care services are especially beneficial in this phase, as they provide professional guidance on symptom management, emotional and spiritual care, and counseling for caregivers. Caregivers should take time to care for themselves by leaning on their support networks, speaking openly about the emotional challenges of this phase, and seeking counseling if needed. By creating a supportive environment, caregivers can provide peace and dignity for their loved ones in this final stage of life.

Q27: What unique challenges do Indigenous communities face in accessing Alzheimer's care, diagnosis, and treatment resources?

Indigenous communities face unique challenges in accessing Alzheimer's care due to geographic isolation, cultural differences, and limited resources. Many communities are located in remote areas, making it difficult to access healthcare services, specialists, and timely diagnosis. Additionally, Western medical approaches to Alzheimer's may not align with Indigenous worldviews, which emphasize holistic, community-based care. This can create barriers to seeking formal treatment, especially if it doesn't reflect cultural values or practices.

Further complicating access is the lack of culturally competent care, with healthcare providers often unfamiliar with Indigenous traditions and needs. There is also a shortage of Indigenous healthcare professionals who can bridge cultural gaps. Compounding these issues is the historical trauma and mistrust of healthcare systems in many Indigenous communities, stemming from past mistreatment. This leads to reluctance in seeking care, particularly when available resources are underfunded or lack cultural sensitivity. Addressing these challenges requires tailored, culturally relevant approaches to care that prioritize community involvement and accessibility.

Q28: What are some strategies for simplifying communication with someone who has Alzheimer's disease?

When communicating with someone who has cognitive challenges, it's helpful to use clear and simple language. Focus on short sentences and one idea at a time. Avoid complex or abstract concepts, as these can increase confusion. It's also important to give them extra time to process the information. Additionally, breaking down tasks or instructions into small, manageable steps can help reduce overwhelm. Using familiar names and concrete references instead of abstract concepts can also ground the conversation and make it easier to follow. Additionally, maintaining a calm, patient demeanor can help to reduce any anxiety the individual may experience while trying to process information.

Q29: How can I ensure the person with Alzheimer's disease is paying attention before I start a conversation?

Gaining the individual's attention before speaking is important. Start by making eye contact and using their name to establish connection. Gentle touch on the arm or a soft gesture can help as well. Reducing distractions, such as turning off the TV or minimizing background noise, is key to helping the person focus. A calm and quiet environment helps reduce confusion and increases the likelihood that the person will engage in the conversation. It may also be helpful to check their emotional state before starting, as stress or anxiety can make it harder for them to focus. If they appear distracted or agitated, try to redirect them gently to regain focus.

Q30: What role does nonverbal communication play when interacting with someone with Alzheimer's disease?

Nonverbal communication becomes essential, especially as verbal skills may decline over time. Gestures, facial expressions, and body language can significantly reinforce verbal messages and help create a supportive environment. Smiling, nodding, and using hand gestures can convey warmth and understanding, even when words are not fully processed. Tone of voice and body language often convey reassurance, which can help maintain connection and understanding. In some cases, touching a hand or gently guiding them can provide additional support and comfort.

Q31: How should I handle moments when the individual with Alzheimer's disease says something that isn' t quite right?

Instead of correcting inaccuracies, it's helpful to validate the individual's emotions and acknowledge their feelings. Correcting them may cause frustration or distress. By using a technique like validation therapy, you can redirect the conversation gently or respond in a way that respects the person's emotional state. If they recall a mistaken memory, it can be more supportive to address the feelings behind that memory rather than correcting the factual details, which can foster a sense of dignity and calm. A simple phrase like "I see how you feel" can help affirm their emotions while steering the conversation to a more positive direction.

Q32: What can I do to make the conversation easier for someone with Alzheimer's disease?

Maintaining a calm and patient demeanor is key. Giving the person ample time to respond without rushing them reduces anxiety and promotes positive communication. If needed, repeating statements or questions can help ensure they understand. Speaking clearly, paired with reassurance and positive reinforcement, can encourage further engagement and make the conversation feel more comfortable. It can also help to limit the number of people in the conversation, as too many voices can cause confusion. Sometimes, simplifying the setting (e.g., sitting face-to-face in a quiet room) can also reduce distractions.

Q33: How can I ensure a comfortable environment for communication?

A quiet, familiar, and well-lit environment is essential for communication. People with cognitive challenges may become overwhelmed by loud noises or visual distractions. Creating a calm, distraction-free space helps the individual focus and feel secure. Warm, soft lighting can make the space feel more inviting, and for those with more advanced cognitive challenges, consistency in the environment can provide a sense of stability. Using familiar objects or pictures in the environment can also offer comfort and help them feel at ease.

Q34: What is the best way to refer to people and places during conversations?

Memory impairments can make it difficult for individuals to recall names, so providing additional context can be helpful. Instead of just using a person's name, try using familiar descriptions, like "your friend from work" or "the neighbor who loves gardening." Photos or visual aids can also support recognition. These simple adjustments make it easier for the individual to understand and engage with the conversation. In some cases, using sensory cues, such as a favorite scent or item of clothing, can help jog their memory and make them feel more comfortable.

Q35: Are there any activities that can help encourage communication with someone with Alzheimer's disease?

Engaging the senses through familiar activities can be a great way to encourage communication. Looking at old photo albums or playing familiar music can help stimulate memory and provide conversation starters. Reminiscence therapy, creative arts, or simple hands-on activities like drawing or crafting can also engage the person and encourage them to express their thoughts and feelings. These activities often tap into long-term memories, which can help prompt meaningful interaction. It's also helpful to allow the person to lead the activity or conversation at their own pace.

Q36: What types of social activities can help stimulate engagement for someone with Alzheimer's disease?

Engaging individuals with Alzheimer's disease in meaningful social activities can provide cognitive and emotional benefits. Activities like group music sessions, gardening, art projects, and reminiscence groups—where people share memories with photo albums or familiar objects—are highly effective. Research suggests these activities help stimulate long-term memory and create a sense of belonging. Keeping sessions short and positive can make it easier for them to participate without feeling overwhelmed. Integrating activities that also engage their senses, like baking or light exercise, can also enhance the overall experience.

Q37: How can caregivers encourage social interaction for a person with Alzheimer's disease who may be withdrawn or hesitant to participate?

Encouraging social engagement often requires gentle guidance. Start by offering activities that align with their past interests or current abilities. Research shows that beginning with small, familiar activities, like looking at a favorite photo album, can help ease them into social settings. Studies highlight that people with Alzheimer's disease often respond well to cues like a friendly tone, calm reassurance, and, when appropriate, a gentle touch to make them feel supported and more willing to engage. Offering choices, such as "Would you like to sing a song or look at a picture book?," can give them a sense of control while easing them into social interaction.

Q38: What are some suitable activities for family gatherings that include a person with Alzheimer's disease?

Family gatherings can be joyful but are best kept low-key to accommodate an individual with Alzheimer's disease. Activities like looking through family photos, singing familiar songs, or enjoying a simple, favorite meal can be comforting and enjoyable. Alzheimer's experts recommend choosing activities that involve familiar sensory elements, as these help individuals feel connected and reduce the likelihood of overstimulation. Keeping interactions short and focused on one or two people at a time can also make the experience positive for everyone involved.

Q39: How can a caregiver balance providing social stimulation with preventing overstimulation for someone with Alzheimer's disease?

Balancing stimulation and calm is key for individuals with Alzheimer's disease. Experts suggest starting with brief social interactions and watching for signs of fatigue, restlessness, or anxiety, which may indicate overstimulation. Quiet environments, consistent routines, and structured activities help create a sense of security. Taking regular breaks and offering a calm space where they can retreat if needed are practical ways to avoid overstimulation. Patience and flexibility are essential to maintaining a positive experience.

Q40: What role can group activities play in supporting socialization and well-being for someone with Alzheimer's disease?

Group activities like memory cafes, art therapy, and gentle movement classes create a supportive social environment where individuals with Alzheimer's disease can engage at their own pace. According to research, group settings offer the benefits of shared connection and cognitive stimulation. Familiar activities in a calm, structured setting can also boost mood, provide comfort, and help reduce feelings of loneliness.

Q41: How can hobbies and interests be adapted into social activities for individuals with Alzheimer's disease?

Adapting previous hobbies is an excellent way to encourage socialization. For instance, if they enjoyed cooking, you could involve them in simple meal prep tasks, such as washing vegetables or stirring. For those who loved gardening, planting seeds or arranging flowers can be calming and enjoyable. Alzheimer's care experts note that simplified versions of past hobbies can evoke happy memories, foster a sense of accomplishment, and keep them connected to familiar interests in a way that supports their current abilities.

Q42: How can I help an individual with Alzheimer's disease feel connected to others when verbal communication is difficult?

When verbal communication is limited, nonverbal methods such as smiling, hand-holding, and maintaining eye contact become essential. Gestures, such as nodding or pointing to objects, also play a role in expressing understanding. Research shows that individuals with Alzheimer's disease continue to respond to emotional cues even as verbal abilities decline, making simple, expressive interactions important. Providing a calm, supportive presence and using visual aids, like photos, can bridge communication gaps and maintain a connection.

Q43: What are some conversation starters that can help engage someone with Alzheimer's in social settings?

Conversation starters should be familiar and evoke positive memories. Try questions like, "What was your favorite hobby?" or "Do you remember your favorite holiday tradition?" Alzheimer's specialists suggest using sensory or visual aids, like family photos or favorite songs, to prompt conversations. Simple questions or reminiscence topics help them feel comfortable and more likely to respond.

Q44: How can social activities help reduce loneliness or anxiety in individuals with Alzheimer's?

Social activities provide emotional and cognitive support, helping reduce feelings of isolation. Alzheimer's research highlights the importance of consistent interaction in reducing anxiety and depression. Even gentle activities, like listening to music or engaging in group crafts, can bring comfort, provide a sense of connection, and reduce agitation. Regular, small-group interactions help establish familiarity and routine, which are beneficial for emotional well-being.

Q45: Are there community-based social activities designed specifically for people with Alzheimer's disease?

Many communities offer programs tailored to individuals with Alzheimer's disease, such as memory cafes, art therapy, and sensory-friendly outings. Memory cafes, in particular, provide a relaxed environment where individuals with Alzheimer's disease and their caregivers can socialize with others. Alzheimer's experts find that these settings, which encourage socialization in a comfortable, structured environment, help promote a sense of normalcy and connection, easing both cognitive and emotional challenges.

Q46: How can technology or virtual gatherings help maintain social connections for people with Alzheimer's disease, especially if in-person is not possible?

Virtual gatherings offer a way for individuals with Alzheimer's disease to connect with loved ones from home, especially when in-person visits are limited. Alzheimer's care providers recommend using simple technology, such as video calls, where a caregiver can help facilitate. Video calls with familiar faces or even watching family videos together can provide comfort and maintain a sense of connection. Familiar routines in these virtual settings can ease feelings of isolation and bring a positive experience.

Q47: What are some low-pressure, quiet social activities that might be comforting for a person with Alzheimer's disease?

Low-stimulation activities, such as looking through photo albums, listening to favorite music, or gentle stretching exercises, are ideal for maintaining calm and comfort. Alzheimer's care guidelines recommend activities that are predictable and familiar, helping individuals feel secure and connected without pressure. These quiet activities can be done with family members or caregivers and encourage meaningful interaction in a soothing way.

Q48: How can I show understanding and empathy when communicating with a loved one who has Alzheimer's disease?

Empathy is key to meaningful communication. When your loved one feels confused or frustrated, try acknowledging their feelings to create a sense of connection. For instance, you might say, "I can see you're feeling frustrated, and that's okay. Let's work through this together." Simple phrases like these show them you understand, which can help ease any distress and make them feel valued.

Q49: How can I help a person with Alzheimer's disease recognize their emotions when they may not be able to express them verbally?

It's common for individuals with Alzheimer's disease to experience difficulty expressing emotions, especially as verbal communication becomes more challenging. Nonverbal cues such as facial expressions, body language, and gestures can provide valuable insight into their emotional state. When they seem agitated or upset, it may help to gently acknowledge their feelings, such as saying, "I can see you're feeling frustrated," and offering comforting actions like holding their hand or using a calming voice. Additionally, using visual aids such as emotion cards or pictures depicting different emotions can encourage the individual to point to or identify how they feel. This can help the caregiver understand and respond more effectively to their emotional needs.

Q50: What non verbal cues can help make communication more effective?

Nonverbal communication—such as smiling, nodding, or offering gentle touches (if the person is comfortable with physical contact)—can significantly enhance the message of warmth and support you're trying to convey. As Alzheimer's disease progresses, a person's verbal abilities may diminish, making it harder for them to fully understand or respond to spoken words. In these cases, gestures and facial expressions become even more important as they can help convey reassurance, affection, and empathy. A simple smile can go a long way in communicating kindness and reducing anxiety, while nodding can signal agreement or acknowledgment, even when words are not fully understood.

When using touch, it's important to ensure that the person feels comfortable and respected. If they are receptive, a gentle touch on the arm or shoulder can provide a sense of closeness and security. Always be mindful of the person's body language to gauge whether touch is welcome, and respect their personal space if they appear uncomfortable.

In addition to these gestures, making eye contact and ensuring that you are positioned at their eye level (without crowding them) can make a significant difference. Facing them directly helps show that you are fully present and engaged, which is especially important when verbal communication may be limited. It creates a sense of connection and conveys that you are listening and paying attention, which can make the person feel valued and understood. A calm and reassuring demeanor, combined with these non verbal cues, can be a powerful tool in helping someone with Alzheimer's feel safe and acknowledged in moments when words alone may not suffice.

Q51: How can individuals with Alzheimer's disease be given choices to help them feel in control?

Offering simple choices, even in small matters, can have a significant impact on someone with Alzheimer's, helping to maintain their sense of independence and dignity. While Alzheimer's may limit decision-making abilities, providing the opportunity to choose—whether it's "Would you like to sit here or over there?" or "Do you want tea or juice?"—can foster a sense of empowerment. Though their ability to make larger decisions may decline, these small, manageable choices allow them to feel that their preferences matter and that they still have some control over their environment. Keeping the options clear and straightforward reduces confusion and ensures that they feel respected and valued.

Additionally, offering choices helps preserve a sense of self, which can be vital in maintaining their emotional well-being. Alzheimer's can sometimes make individuals feel disconnected from their previous life, so giving them the ability to make even small decisions reinforces their identity and role in daily activities. By providing clear, simple options, you help reduce feelings of frustration and empower them to actively participate in their routine. This practice not only enhances their autonomy but also improves their overall quality of life, promoting a sense of belonging and involvement in decisions, no matter how small.

Q52: What are some tips for actively listening to someone with Alzheimer's disease?

Active listening is a cornerstone of effective communication, particularly for someone with Alzheimer's. As their ability to express themselves may be affected, it's important to give your loved one extra time to respond without rushing or interrupting them. This patience allows them the space to gather their thoughts and communicate at their own pace. Interrupting or finishing their sentences may unintentionally convey impatience, making them feel less confident or discouraged. Instead, provide them with a calm, supportive environment where they can take their time, knowing you're there to listen without judgement or pressure.

In addition to giving them time to speak, maintaining eye contact and offering non verbal cues—like nodding or leaning in slightly—reinforces that you are truly engaged and focused on what they're saying. Even if there are long pauses, patiently waiting shows respect and encourages them to continue sharing. Active listening also reduces the stress or frustration that can arise from misunderstandings or a rushed conversation. By allowing them the opportunity to express themselves fully, you help ensure that they feel understood and valued, fostering a deeper emotional connection and improving the overall quality of your interactions.

Q53: How can I use memories to engage individuals with Alzheimer's disease in conversation?

Talking about positive memories is a wonderful way to foster connection with individuals with Alzheimer's disease. Sharing stories about the past can help bring comfort and reassurance, especially when Alzheimer's disease affects their short-term memory. Asking about their favorite childhood memories, special occasions, or meaningful experiences allows them to revisit moments that made them happy. You can also use photos, familiar music, or cherished objects as conversation starters to trigger these memories. These tangible items often serve as emotional anchors, allowing your loved one to reconnect with a time they remember well, even if they can no longer recall recent events.

Reminiscing about positive memories can have therapeutic effects, not only strengthening your emotional bond but also helping the individual feel more grounded. It can offer a sense of familiarity and comfort, reducing feelings of confusion or anxiety. Moreover, discussing happy memories creates opportunities for joy, laughter, and meaningful exchanges. It reminds them that despite the challenges posed by Alzheimer's, they still have a rich history and identity worth celebrating.

Bibliography

1. ***Wang, H.-X., Karp, A., Winblad, B., & Fratiglioni, L. (2002).*** *Late-life engagement in social and leisure activities is associated with a decreased risk of dementia: A longitudinal study from the Kungsholmen project. American Journal of Epidemiology, 155(12), 1081–1087. https://doi.org/10.1093/aje/kwf063*
2. ***Ellwardt, L., Aartsen, M., Deeg, D., & Steverink, N. (2013).*** *Does loneliness mediate the relation between social support and cognitive functioning in later life? Social Science & Medicine, 98, 116–124. https://doi.org/10.1016/j.socscimed.2013.08.016*

3. **Seeman, T. E., Lusignolo, T. M., Albert, M., & Berkman, L. (2001).** *Social relationships, social support, and patterns of cognitive aging in healthy, high-functioning older adults: MacArthur studies of successful aging. Health Psychology, 20(4), 243–255. https://doi.org/10.1037/0278-6133.20.4.243*
4. **Alzheimer's Association. (2023).** *Alzheimer's disease facts and figures. Alzheimer's & Dementia, 19(3), 391–460. https://doi.org/10.1002/alz.12638*
5. **Mitchell, S. L., Teno, J. M., Kiely, D. K., Shaffer, M. L., Jones, R. N., Prigerson, H. G., et al. (2009).** *The clinical course of advanced dementia. New England Journal of Medicine, 361(16), 1529–1538. https://doi.org/10.1056/NEJMoa0902234*
6. **Vick, J. B., Amjad, H., Smith, K. C., Boyd, C. M., Gitlin, L. N., Roth, D. L., et al. (2018).** *Patient and caregiver perspectives on managing multiple health conditions in dementia. Journal of the American Geriatrics Society, 66(10), 1887–1895. https://doi.org/10.1111/jgs.15453*
7. **Alzheimer's Association. (2023).** *Communication and Alzheimer's. Alzheimer's Association.* https://www.alz.org/professionals/professional-providers/communication-and-alzheimers
8. **Small, J. A., Gutman, G., Makela, S., & Hillhouse, B. (2003).** *Effectiveness of communication strategies used by caregivers of persons with Alzheimer's disease during activities of daily living. Journal of Speech, Language, and Hearing Research, 46(2), 353–367. https://doi.org/10.1044/1092-4388(2003/027)*

CHAPTER 9

Legal and Financial Planning for Alzheimer's Disease

Legal and financial planning is a critical component of managing Alzheimer's disease, especially as cognitive decline progresses. The complexities of Alzheimer's require families to take proactive steps in safeguarding the individual's finances, ensuring access to necessary resources, and preparing for long-term care. Early and thoughtful planning can help ease the burdens on families and honor the wishes of those affected by the disease.

This chapter provides insights into essential legal documents, such as powers of attorney and advance directives, that empower families to make informed decisions. We also discuss strategies for protecting assets, navigating the costs of care, and understanding the legal responsibilities of caregivers. By addressing these important considerations, families can better prepare for the future while maintaining dignity and security for their loved ones with Alzheimer's.

Q1: What legal documents should individuals with Alzheimer's disease and their families consider creating?

Individuals with Alzheimer's disease and their families should consider creating legal documents such as a durable power of attorney for finances and healthcare, a living will or advance directive, a will or trust, and guardianship or conservatorship arrangements, if necessary. These documents can help ensure that the individual's wishes are respected and that their affairs are managed effectively as the disease progresses.

Q2: How can families plan for the financial costs associated with Alzheimer's disease care?

Families can plan for the financial costs associated with Alzheimer's disease care by assessing their current financial situation, exploring long-term care insurance options, researching government assistance programs such as Medicaid, creating a budget for caregiving expenses, exploring community resources and support services, and consulting with financial advisors or elder law attorneys for guidance.

Q3: What are the potential risks of financial exploitation or fraud for individuals with Alzheimer's disease?

Individuals with Alzheimer's disease may be at risk of financial exploitation or fraud due to cognitive impairments, vulnerability to manipulation, and difficulty managing finances independently. Family members and caregivers should monitor financial transactions, safeguard important documents and valuables, set up safeguards such as automatic bill payments and direct deposits, and educate the individual about common scams and fraud schemes.

Q4: What legal and financial resources are available to individuals and families affected by Alzheimer's disease?

Legal and financial resources available to individuals and families affected by Alzheimer's disease include legal aid organizations, elder law attorneys, financial planners specializing in eldercare, government agencies such as the Social Security Administration and Department of Veterans Affairs, Alzheimer's associations offering legal and financial planning workshops, and online resources and toolkits for legal and financial planning.

Q5: How can individuals with Alzheimer's disease ensure their wishes are upheld regarding medical decisions and financial matters?

In order to ensure their wishes are upheld regarding medical decisions and financial matters, individuals with Alzheimer's disease should first create advance directives such as a living will or advance directive. These documents outline their preferences for medical treatment and end-of-life care, providing clear guidance to healthcare providers and family members. Designating a durable power of attorney for finances allows them to appoint a trusted person to manage financial affairs according to their wishes. Keeping all legal documents, including wills or trusts, updated and communicating openly with family, caregivers, and healthcare providers are crucial steps. Regular reviews of these documents and seeking advice from legal and financial professionals specialized in elder care ensure comprehensive planning that aligns with their evolving needs and desires throughout the progression of the disease.

Q6: What steps can families take to protect assets and plan for long-term care needs as Alzheimer's disease progresses?

Families can protect assets and plan for long-term care needs as Alzheimer's disease progresses by first conducting a comprehensive financial assessment to understand available resources. They should consider purchasing long-term care insurance early to cover future care costs. Essential legal documents, such as durable powers of attorney and advance directives, should be established to designate decision-makers and outline care preferences. Asset protection strategies, like setting up trusts, can safeguard assets, while understanding Medicaid eligibility and planning ahead can help cover potential long-term care expenses. Effective healthcare coordination, family communication, and seeking guidance from financial advisors and elder law attorneys ensure a well-rounded approach to financial and care planning, providing stability and support throughout the disease's progression.

Q7: How do legal and financial planning considerations differ for early-stage versus late-stage Alzheimer's disease?

Legal and financial planning considerations differ significantly between early-stage and late-stage Alzheimer's disease. In the early stages, individuals can typically participate in decision-making and should focus on establishing durable powers of attorney for finances and healthcare, updating legal documents, and exploring long-term care insurance options. As Alzheimer's progresses to later stages and cognitive abilities decline, caregivers or designated agents must manage financial and healthcare decisions through established legal powers. Planning for Medicaid eligibility, considering guardianship or conservatorship if necessary, and maintaining open communication with family and professional advisors become crucial to ensure that the individual's wishes are upheld and their financial security is protected throughout the progression of the disease.

Q8: What role does guardianship or conservatorship play in managing the affairs of someone with advanced Alzheimer's disease?

Guardianship or conservatorship plays a critical role in managing the affairs of someone with advanced Alzheimer's disease when they are no longer capable of making decisions independently. In these cases, a court appoints a guardian or conservator to act on behalf of the individual, ensuring their welfare and managing their financial and personal affairs. This legal arrangement allows the appointed individual to make decisions regarding healthcare, living arrangements, financial management, and other necessary aspects of life, ensuring that the person with Alzheimer's disease receives proper care and support as their cognitive abilities decline. Guardians or conservators are typically responsible for advocating for the individual's best interests, coordinating medical care, managing finances, and ensuring that legal obligations are met in accordance with the law and the wishes outlined in legal documents, such as advance directives or powers of attorney.

Q9: How can individuals and families find reputable financial advisors or elder law attorneys experienced in Alzheimer's related planning?

To find reputable financial advisors or elder law attorneys experienced in Alzheimer's related planning, individuals and families can start by seeking referrals from trusted sources such as family members, healthcare providers, or local Alzheimer's associations. They should look for professionals who are members of recognized organizations like the National Academy of Elder Law

Attorneys (NAELA) or the Financial Planning Association (FPA), as these groups often have directories of qualified experts. Online research can provide additional insights into the professionals' backgrounds and specialties, while scheduling consultations allows families to discuss specific needs, such as Medicaid planning, long-term care insurance, and estate management tailored to Alzheimer's disease. Evaluating credentials, experience, and client testimonials ensures selecting professionals who can provide effective guidance and support through the complexities of Alzheimer's related legal and financial planning.

Q10: What are the benefits of creating a trust versus a will for managing assets and ensuring their distribution according to wishes?

Creating a trust offers several benefits over a will for managing assets and ensuring their distribution according to wishes, especially in the context of Alzheimer's disease. Trusts provide greater privacy and efficiency by avoiding probate, ensuring that asset management and distribution can occur privately and swiftly. They also offer flexibility in asset management during incapacity, allowing for seamless transitions of responsibility through designated trustees. Trusts can be tailored with specific instructions for asset distribution, offering more control and protection for beneficiaries, and potentially reducing costs associated with probate. Overall, trusts provide a comprehensive solution for managing assets both before and after death, making them a valuable option for individuals planning for Alzheimer's related challenges.

Q11: How does the process of applying for Medicaid differ for individuals with Alzheimer's disease compared to other conditions?

The process of applying for Medicaid can differ for individuals with Alzheimer's disease compared to those with other conditions primarily due to the cognitive impairments associated with Alzheimer's. Individuals with Alzheimer's may face challenges in understanding and completing the application process independently, requiring assistance from caregivers or appointed representatives. Documentation of medical and financial history must often account for the progressive nature of Alzheimer's, including detailed records of medical diagnoses, functional impairments, and financial resources. Moreover,

Medicaid applications may need additional advocacy and legal support to navigate complexities and ensure eligibility criteria are met amidst cognitive decline.

Q12: What are the tax implications for families caring for a loved one with Alzheimer's disease, and how can they minimize tax burdens?

Families caring for a loved one with Alzheimer's disease should be aware of several tax implications and strategies to minimize their tax burdens. They can potentially deduct qualified medical expenses, including those related to Alzheimer's care, if they exceed a certain percentage of their adjusted gross income (AGI). Utilizing flexible spending accounts (FSAs) or health savings accounts (HSAs) for medical expenses can also reduce taxable income. Additionally, the dependent care tax credit may apply if the individual with Alzheimer's qualifies as a dependent. Careful record-keeping of all expenses related to caregiving is essential, and consulting with tax professionals specializing in eldercare can help navigate available deductions and credits effectively.

Q13: How can families prepare financially for potential home modifications or long-term care facility costs associated with Alzheimer's care?

Families preparing for potential home modifications or long-term care facility costs associated with Alzheimer's care should start by assessing their current financial situation and creating a detailed budget. This includes identifying ongoing caregiving expenses and considering long-term care insurance options to cover additional costs not covered by health insurance or Medicare. Planning for home modifications, such as safety enhancements, and researching local long-term care facilities are crucial steps. Updating legal documents and consulting with financial advisors can provide further guidance on managing expenses, maximizing resources, and ensuring financial readiness for Alzheimer's care needs.

Q14: What are the legal responsibilities and considerations for caregivers managing finances on behalf of someone with Alzheimer's disease?

Caregivers managing finances for someone with Alzheimer's disease shoulder significant legal responsibilities. They typically do so through a durable power of attorney, granting them authority over banking, investments, and property decisions. This role demands adherence to fiduciary duties, ensuring all actions benefit the individual's best interests and are meticulously documented. In cases where a power of attorney isn't in place or sufficient, obtaining conservatorship or guardianship may be necessary through legal channels. Compliance with tax obligations, transparent communication with family, and seeking guidance from legal professionals specializing in elder law are crucial for navigating these complex responsibilities effectively.

Q15: How can families ensure continuity of care and financial stability for the person with Alzheimer's disease as their condition progresses?

Families can ensure continuity of care and financial stability for a person with Alzheimer's disease by establishing a thorough long-term care plan that encompasses legal, financial, and medical considerations. This includes updating legal documents like powers of attorney and wills, seeking guidance from financial advisors on managing assets and planning for long-term care costs, and coordinating care among healthcare providers and support networks. Regular monitoring and adjustment of the care plan based on the individual's changing needs, along with staying connected to community resources for additional support, are essential to maintaining quality care and stability as the disease progresses.

Q16: What are the ethical considerations involved in managing financial and medical decisions for someone with Alzheimer's disease?

Managing financial and medical decisions for someone with Alzheimer's entails delicate ethical considerations, especially as cognitive impairments progress. A key ethical challenge is balancing respect for the individual's autonomy with the principle of beneficence, which involves acting in their best interest. In the early stages, it's important to honor their right to self-determination by involving them in decision-making whenever possible. According to research, using " substituted judgment "—making decisions based on what the person would have chosen based on their values—can help maintain their dignity. However, as cognitive decline advances, decisions may require a greater focus on safety and welfare, sometimes limiting autonomy to ensure well-being. Consulting healthcare professionals and ethics boards when faced with difficult choices can ensure these decisions align with both ethical guidelines and the individual's prior wishes.

Q17: What impact does early financial and legal planning have on the quality of life for individuals with Alzheimer's and their caregivers?

Early financial and legal planning can significantly improve the quality of life for individuals with Alzheimer's and their caregivers by offering a structured approach to future care needs. Research suggests that proactive planning alleviates stress for caregivers, reduces the risk of financial hardships, and provides clear direction for medical and financial decisions as the disease progresses. Establishing documents such as durable powers of attorney and advance healthcare directives early on ensures that the individual's wishes are respected, which also helps reduce family conflicts. Furthermore, early planning enables families to access benefits and resources, allowing caregivers to focus on providing emotional support rather than dealing with crises.

Q18: What are the benefits and limitations of using long-term care insurance in Alzheimer's financial planning?

Long-term care insurance can provide critical support for Alzheimer's care, covering expenses such as assisted living, in-home care, and memory care facilities, which may otherwise strain a family's finances. Research found that long-term care insurance helps families manage these high costs without depleting personal assets, offering peace of mind. However, there are limitations: premiums can be prohibitively expensive, particularly for those who purchase policies later in life or have pre-existing conditions. Additionally, many policies contain exclusions or benefit caps that may not fully cover advanced Alzheimer's care needs, requiring families to supplement with personal funds. Therefore, purchasing long-term care insurance early and thoroughly understanding policy terms is essential for effective planning.

Q19: What are the legal implications of sharing caregiving responsibilities among multiple family members?

When caregiving duties are shared among family members, clear legal arrangements and communication are essential to avoid conflicts and ensure effective care. Documents such as a durable power of attorney and trust agreements clarify each person's role in financial and healthcare decisions, minimizing the risk of misunderstandings and disputes. It is suggested that holding regular family meetings and involving an elder law mediator can help define each family member's responsibilities and avoid future conflicts. Documenting caregiving roles and maintaining transparency about financial decisions are crucial steps for preventing allegations of mismanagement and ensuring that the individual with Alzheimer's receives consistent support.

Q20: What are the tax benefits and deductions available for families supporting a relative with Alzheimer's, and how can these be accessed?

Families supporting a loved one with Alzheimer's may qualify for several tax benefits and deductions. For example, they can deduct medical expenses related to Alzheimer's care that exceed a certain percentage of their adjusted gross income (AGI). Additional tax advantages include dependent care credits if the person with Alzheimer's qualifies as a dependent. To access these benefits, caregivers should meticulously document all caregiving-related expenses, such as home modifications and medical bills. Consulting a tax professional experienced in elder care can ensure families maximize available deductions, which can substantially ease the financial burden of caregiving.

Bibliography

1. ***Alzheimer's Association. (2020).*** *2020 Alzheimer's disease facts and figures. Alzheimer's & Dementia, 16(3), 391–460. https://doi.org/10.1002/alz.12068*
2. ***Arribas-Ayllon, M. (2018).*** *Dementia, vulnerability and well-being: A qualitative study of policy and practice in care homes. BMC Geriatrics, 18(1), 6. https://doi.org/10.1186/s12877-018-0714-5*
3. ***Karlawish, J. (2017).*** *Addressing the ethical, policy, and social challenges of preclinical Alzheimer disease. Neurology, 77(15), 1487–1493. https://doi.org/10.1212/WNL.0000000000003561*
4. ***Knapp, M. (2020).*** *Dementia, social work, and research: The intersection of policy and practice. Social Work in Health Care, 45(1), 15–34. https://doi.org/10.1080/00981389.2020.1692682*
5. ***McGarry, B. E., & Temkin-Greener, H. (2020).*** *The impact of Medicaid home and community-based services on Alzheimer's disease and related dementias. Journal of Aging and Health, 32(4–5), 358–375. https://doi.org/10.1177/0898264319860987*
6. ***Mitchell, S. L., Teno, J. M., Kiely, D. K., Shaffer, M. L., Jones, R. N., Prigerson, H. G., et al. (2009).*** *The clinical course of advanced dementia. New England Journal of Medicine, 361(16), 1529–1538. https://doi.org/10.1056/NEJMoa0902234*
7. ***Robinson, L., Tang, E., & Taylor, J. P. (2015).*** *Dementia: Timely diagnosis and early intervention. BMJ, 350, h3029. https://doi.org/10.1136/bmj.h3029*

8. **Shanahan, M. J., & White-Means, S. I. (2018).** *The tax treatment of caregiving expenses for Alzheimer's disease and other long-term care. Journal of Aging & Social Policy, 30(5), 497–510. https://doi.org/10.1080/08959420.2018.1511977*
9. **Sörensen, S., & Conwell, Y. (2019).** *Issues in dementia caregiving: Effects on mental and physical health, intervention strategies, and research needs. American Journal of Geriatric Psychiatry, 19(6), 491–496. https://doi.org/10.1097/JGP.0b013e3181f0f0c3*
10. **Vick, J. B., Amjad, H., Smith, K. C., Boyd, C. M., Gitlin, L. N., Roth, D. L., et al. (2018).** *Patient and caregiver perspectives on managing multiple health conditions in dementia. Journal of the American Geriatrics Society, 66(10), 1887–1895. https://doi.org/10.1111/jgs.15453*
11. **Yaffe, K. (2020).** *Patient and caregiver characteristics and nursing home placement in patients with dementia. JAMA, 287(16), 2090–2097. https://doi.org/10.1001/jama.2020.5805*

CHAPTER 10

Research and Advances in Alzheimer's Disease

The scientific community continues to make strides in understanding Alzheimer's disease, aiming to uncover more effective diagnostic tools, treatments, and prevention strategies. From advancements in biomarkers and genetic research to lifestyle interventions and precision medicine, Alzheimer's research offers hope for better outcomes and improved quality of life for those affected.

In this chapter, we explore the latest findings in Alzheimer's disease research, including emerging therapies, the role of genetic risk factors, and the impact of inflammation on disease progression. Additionally, we examine the potential of lifestyle modifications and innovative diagnostic tools to support early detection and personalized treatment plans. By staying informed about these advances, individuals, families, and caregivers can make empowered choices grounded in cutting-edge science and compassionate care.

–

Q1: What are the latest research findings and advances in Alzheimer's disease treatment and prevention?

Recent research in Alzheimer's disease has focused on developing disease-modifying treatments targeting amyloid beta and tau proteins, exploring immunotherapy and gene therapy approaches, investigating lifestyle interventions and biomarkers for early detection and prevention, and exploring precision medicine strategies based on individualized risk factors and genetic profiles.

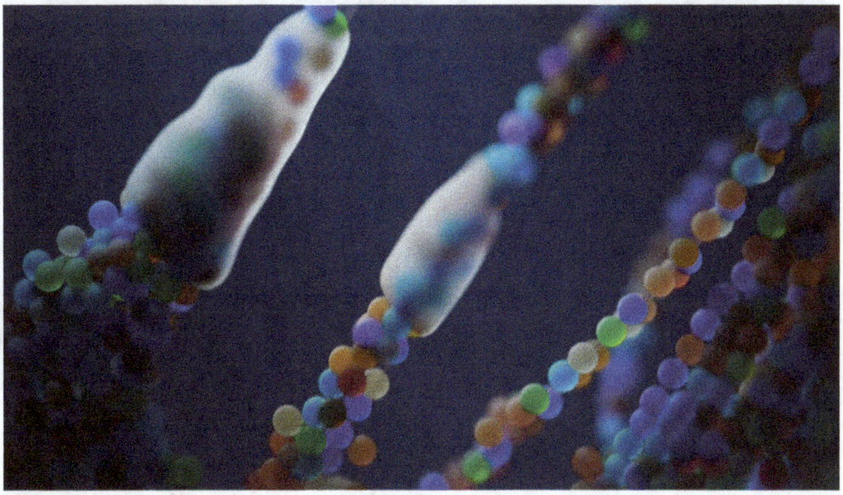

▲ **FIGURE 10.1.**
Close-up of proteins.

Q2: How can individuals participate in Alzheimer's disease research studies and clinical trials?

Individuals interested in participating in Alzheimer's disease research studies and clinical trials can explore opportunities through research centers, academic institutions, Alzheimer's associations, and online databases such as ClinicalTrials.gov. Participation may involve screening assessments, medical

▲ **FIGURE 10.2.**
A happy family.

evaluations, and informed consent procedures, and individuals should carefully consider the risks and benefits before enrolling.

Q3: What are the challenges and future directions of Alzheimer's disease research?

Challenges in Alzheimer's disease research include identifying effective biomarkers for early diagnosis and intervention, developing treatments that target multiple pathological processes, addressing heterogeneity in disease presentation and progression, improving clinical trial design and recruitment strategies, and promoting interdisciplinary collaboration and data sharing across research communities.

Q4: How can individuals stay informed about the latest research developments and breakthroughs in Alzheimer's disease?

Individuals can stay informed about the latest research developments and breakthroughs in Alzheimer's disease by subscribing to research newsletters, attending scientific conferences and symposiums, following reputable research institutions and organizations on social media, participating in community educational events, and accessing online resources such as research blogs, podcasts, and webinars.

Q5: What role do genetics play in Alzheimer's disease risk and progression?

Genetics play a significant role in Alzheimer's disease risk and progression, with certain genetic variants influencing susceptibility to the condition. The APOE gene, in particular, has been implicated as a major genetic risk factor, with the APOE4 allele associated with increased risk and earlier onset of Alzheimer's. Other genetic factors, including mutations in genes such as APP, PSEN1, and PSEN2, can also contribute to familial forms of the disease.

Q6: How does inflammation contribute to Alzheimer's disease pathology?

Inflammation plays a complex role in Alzheimer's disease pathology, with both beneficial and detrimental effects on brain health. While acute inflammation is a normal response to injury or infection and can help clear harmful substances from the brain, chronic inflammation can contribute to neuronal damage, synaptic dysfunction, and neurodegeneration in Alzheimer's. Understanding the mechanisms of inflammation in the brain may lead to the development of novel anti-inflammatory therapies for the disease.

Q7: What are the potential benefits of lifestyle interventions in Alzheimer's disease prevention and management?

Lifestyle interventions, including regular physical exercise, healthy diet, cognitive stimulation, social engagement, and adequate sleep, have been shown to have potential benefits in Alzheimer's disease prevention and management. These interventions can promote brain health, reduce risk factors for cognitive decline, and enhance overall well-being in individuals at risk for or living with Alzheimer's.

Q8: How do sleep disturbances impact Alzheimer's disease risk and progression?

Sleep disturbances, including insomnia, sleep apnea, and disruptions in the sleep–wake cycle, have been associated with an increased risk of Alzheimer's disease and may contribute to disease progression. Sleep plays a crucial role in brain function, including memory consolidation, toxin clearance, and synaptic plasticity, so disturbances in sleep patterns may exacerbate underlying neuropathological changes in Alzheimer's.

Q9: What are the challenges of diagnosing Alzheimer's disease in its early stages?

Diagnosing Alzheimer's disease in its early stages presents several challenges, including overlapping symptoms with normal aging and other neurodegenerative disorders, limited diagnostic tools for detecting early pathology, variability in disease presentation and progression, and the need for comprehensive assessments combining clinical, cognitive, imaging, and biomarker data.

Q10: How are biomarkers used in Alzheimer's disease research and diagnosis?

Biomarkers, including imaging markers such as amyloid PET scans and MRI, cerebrospinal fluid (CSF) markers such as amyloid beta and tau proteins, and blood-based markers, are used in Alzheimer's disease research and diagnosis to detect underlying pathology, track disease progression, and monitor treatment response. Biomarkers may help identify individuals at risk for Alzheimer's, facilitate early diagnosis, and inform the development of targeted therapies.

Q11: What are the current challenges in developing effective treatments for Alzheimer's disease?

Developing effective treatments for Alzheimer's disease faces several challenges, including the complexity of the disease pathology, the need for interventions targeting multiple pathological processes, difficulties in translating preclinical findings to clinical efficacy, heterogeneity in disease presentation and progression, and the lack of validated biomarkers for patient stratification and treatment monitoring.

Q12: What are the potential roles of gut microbiota in Alzheimer's disease pathogenesis?

Emerging evidence suggests that the gut microbiota, the diverse community of microorganisms residing in the gastrointestinal tract, may play a role in Alzheimer's disease pathogenesis. Dysbiosis, or imbalance in the gut microbiota composition, has been associated with neuroinflammation, amyloid deposition, and cognitive impairment in Alzheimer's, highlighting the potential for targeting the gut– brain axis as a therapeutic strategy.

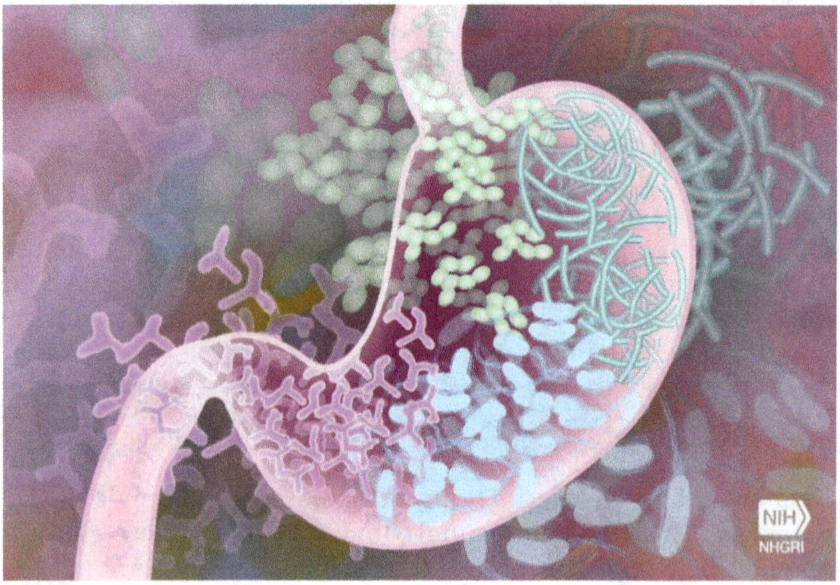

▲ **FIGURE 10.3.**
Beneficial gut bacteria.

Q13: How does vascular health impact Alzheimer's disease risk and progression?

Vascular health, including factors such as hypertension, diabetes, hypercholesterolemia, and smoking, can impact Alzheimer's disease risk and progression. Vascular risk factors contribute to vascular pathology, including cerebral small vessel disease and ischemic injury, which may interact with Alzheimer's pathology to accelerate cognitive decline and increase the risk of dementia.

Q14: What are the potential benefits of non-pharmacological interventions in Alzheimer's disease management?

Non-pharmacological interventions, including cognitive training, reminiscence therapy, music therapy, art therapy, and multisensory stimulation, have been shown to have potential benefits in Alzheimer's disease management. These interventions can improve cognitive function, mood, behavior, and quality of life for individuals living with Alzheimer's, often complementing pharmacological treatments.

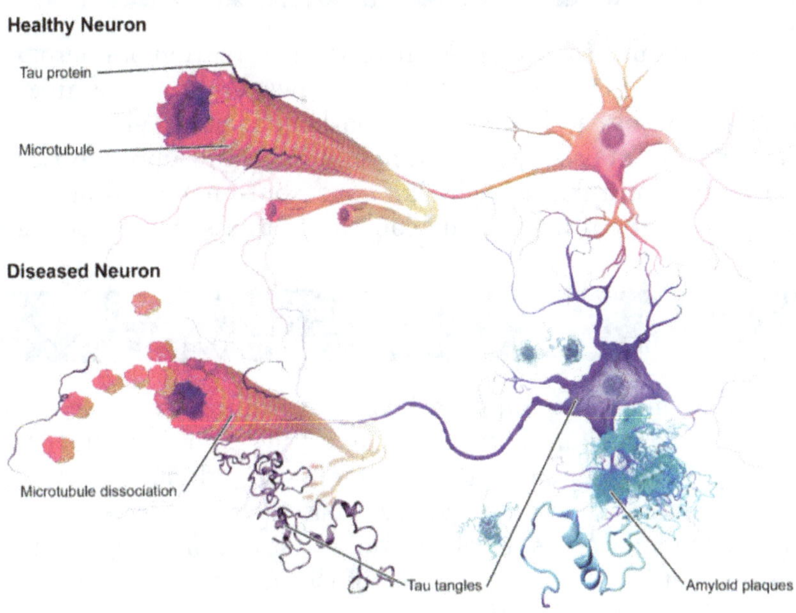

▲ **FIGURE 10.4.**
Healthy vs. unhealthy neurons.

Q15: How do environmental factors such as air pollution and exposure to toxins impact Alzheimer's disease risk?

Environmental factors such as air pollution, heavy metals, pesticides, and industrial toxins have been implicated in Alzheimer's disease risk. These environmental exposures can contribute to neuroinflammation, oxidative stress, and neurodegeneration, potentially increasing susceptibility to Alzheimer's and other neurodegenerative disorders.

Q16: What are the potential mechanisms underlying synaptic dysfunction in Alzheimer's disease?

Synaptic dysfunction, characterized by impairments in synaptic plasticity, neurotransmission, and synaptic connectivity, is a hallmark of Alzheimer's disease pathology. Mechanisms underlying synaptic dysfunction include the accumulation of amyloid beta and tau proteins, disruptions in calcium homeostasis, mitochondrial dysfunction, neuroinflammation, and alterations in neurotrophic factors, highlighting potential targets for therapeutic intervention.

Q17: How do neuroinflammatory processes contribute to Alzheimer's disease pathogenesis?

Neuroinflammatory processes, involving activation of microglia, astrocytes, and cytokine signaling pathways, play a significant role in Alzheimer's disease pathogenesis. While acute inflammation may help clear amyloid beta and promote tissue repair, chronic inflammation can exacerbate neuronal damage, disrupt synaptic function, and contribute to disease progression in Alzheimer's.

Q18: What are the potential benefits of precision medicine approaches in Alzheimer's disease treatment?

Precision medicine approaches, leveraging advances in genetics, biomarkers, and data analytics, hold promise for tailoring Alzheimer's disease treatments to individual patients' characteristics, including genetic risk factors, disease stage, and molecular profiles. Personalized interventions may improve treatment efficacy, reduce adverse effects, and optimize outcomes for individuals living with Alzheimer's.

Q19: How do sex and gender differences influence Alzheimer's disease risk and progression?

Sex and gender differences influence Alzheimer's disease risk and progression, with women being disproportionately affected by the condition. Hormonal factors, genetic predisposition, lifestyle factors, and social determinants of health contribute to sex differences in Alzheimer's risk, highlighting the importance of considering gender-specific approaches to prevention, diagnosis, and treatment.

Q20: What advancements have been made in blood-based biomarkers for Alzheimer's disease diagnosis?

Recent advancements in blood-based biomarkers for Alzheimer's disease are revolutionizing diagnostic approaches by providing more accessible, cost-effective, and minimally invasive testing options. Traditional diagnostic methods, such as cerebrospinal fluid (CSF) analysis and PET imaging, though accurate, are invasive and costly, often limiting their use to specialized centers. Blood-based markers, including plasma amyloid beta, phosphorylated tau (p-tau), and neurofilament light (NFL) protein, correlate with established Alzheimer's pathology and cognitive decline, allowing for early and relatively easy detection of disease. These advancements are particularly promising for broader, early-stage screening in primary care settings. However, ongoing research aims to address challenges in assay standardization, clinical accuracy, and the validation of biomarkers across diverse populations before these tests can become a routine part of clinical practice.

Q21: What are the potential benefits of immunotherapy in Alzheimer's disease treatment?

Immunotherapy represents a groundbreaking approach to Alzheimer's treatment, focusing on the removal of amyloid beta and tau proteins, which accumulate abnormally in the brain. Monoclonal antibodies, such as aducanumab, target amyloid plaques and have shown promise in slowing cognitive decline by reducing amyloid deposits in clinical trial participants. Immunotherapy's potential benefits extend beyond plaque reduction, as it offers a disease-modifying approach rather than merely alleviating symptoms. However, these therapies are not without challenges. Side effects, including amyloid-related imaging abnormalities (ARIA), require careful patient monitoring, and further studies are needed to evaluate long-term safety and effectiveness across diverse patient groups. Continued research aims to refine immunotherapy protocols, maximize patient benefit, and expand access to those most likely to benefit from these treatments.

Q22: How can machine learning enhance Alzheimer's disease research and diagnosis?

Machine learning (ML) is reshaping Alzheimer's research by processing large, complex datasets, and integrating information from neuroimaging, genetic profiles, cognitive testing, and other clinical assessments. ML algorithms are capable of identifying patterns in these data that may indicate early Alzheimer's pathology, improve diagnostic precision, and predict disease progression. ML has been instrumental in developing risk assessment tools and predictive models that allow researchers and clinicians to stratify patients for targeted interventions. Despite these advancements, ML applications face limitations, including the need for high-quality, diverse data to ensure model accuracy and generalizability. Furthermore, the interpretability of ML models remains a challenge, as many algorithms function as " black boxes " with complex decision-making processes that are difficult to understand. Addressing these issues through ongoing research and collaboration will be essential for ML to reach its full potential in clinical settings.

Q23: What are the latest findings on the role of tau protein in Alzheimer's disease progression?

Tau protein is increasingly recognized as a key player in Alzheimer's disease progression, with tau tangles spreading in the brain in a manner closely correlated with disease severity. Unlike amyloid beta, which accumulates early in the disease, tau tangles are more closely tied to neurodegeneration and cognitive decline. Research has shown that tau spreads along connected neural pathways, progressively impacting areas responsible for memory, cognition, and executive functions. Given its central role in disease progression, tau has become a primary target for therapeutic development. Strategies such as tau phosphorylation inhibitors and tau-directed immunotherapy aim to disrupt tau aggregation, slow disease spread, and protect neurons. These findings underscore the importance of tau in Alzheimer's pathology and offer new avenues for disease-modifying treatments that could be more effective than amyloid-targeted therapies alone.

Q24: What progress has been made in understanding the role of gut microbiota in Alzheimer's disease?

The gut–brain axis has revealed that gut microbiota composition may influence Alzheimer's disease pathology, highlighting an unexpected link between digestive and brain health. Studies have found that dysbiosis, or an imbalance in gut microbiota, can trigger neuroinflammation, promote amyloid beta accumulation, and impair cognitive function. These effects are believed to arise from pro-inflammatory molecules produced by certain gut bacteria that can travel to the brain and impact neuronal health. Potential interventions, including probiotics, dietary modifications, and microbiota-targeting drugs, are under investigation as therapeutic strategies for Alzheimer's prevention and management. This emerging research suggests that maintaining a balanced gut microbiota could reduce Alzheimer's risk or slow disease progression, opening up novel approaches to treatment beyond the brain.

Q25: What are the potential roles of mitochondrial dysfunction in Alzheimer's disease pathogenesis?

Mitochondrial dysfunction has been implicated as a critical factor in Alzheimer's disease, as it contributes to neuronal energy deficits, oxidative stress, and ultimately cell death. Mitochondria, often referred to as the "powerhouses of the cell," provide energy for cellular processes. In Alzheimer's, mitochondrial dysfunction leads to reduced energy production and increased oxidative stress, which can further damage neurons and promote amyloid beta and tau accumulation. Experimental therapies focused on enhancing mitochondrial function—such as the use of antioxidants, coenzyme Q10, and mitochondrial-targeting compounds—are being explored as potential interventions to mitigate neurodegeneration and slow Alzheimer's progression. While promising, these treatments require further research to evaluate their efficacy in clinical trials and determine how they can best support neuronal health.

Q26: How does metabolic health impact Alzheimer's disease risk and progression?

Metabolic health, particularly conditions like insulin resistance, obesity, and dyslipidemia, has been increasingly linked to Alzheimer's disease risk and progression. Insulin resistance in the brain, commonly associated with type 2 diabetes, can disrupt glucose metabolism, exacerbating amyloid plaque formation and promoting neurodegeneration. These findings have sparked interest in investigating Alzheimer's as a potential "type 3 diabetes," where impaired insulin signaling in the brain contributes to cognitive decline. Lifestyle modifications, such as a healthy diet and regular exercise, have shown potential in reducing Alzheimer's risk by improving metabolic health. Additionally, diabetes medications, including metformin, are being studied for their neuroprotective effects, indicating that metabolic health management could play a central role in both Alzheimer's prevention and treatment.

Q27: What impact does the APOE genotype have on Alzheimer's disease treatment response?

The APOE genotype, particularly the APOE4 allele, is a well-established genetic risk factor for Alzheimer's disease, affecting both disease risk and age of onset. Research has shown that individuals with the APOE4 allele may respond differently to Alzheimer's treatments, particularly amyloid-targeting therapies. For example, APOE4 carriers may experience varying levels of amyloid plaque reduction or different side effect profiles, making it crucial to tailor treatments accordingly. Precision medicine approaches aim to incorporate genetic information like APOE genotype into personalized treatment plans, optimizing therapeutic efficacy while minimizing adverse effects. This approach underscores the potential for genetic information to guide more effective, individualized interventions, a promising step toward more targeted Alzheimer's care.

Bibliography

1. ***Arnold, S. E., Arvanitakis, Z., Macauley-Rambach, S. L., et al. (2018).*** *Brain insulin resistance in type 2 diabetes and Alzheimer's disease: Concepts and conundrums. Nature Reviews Neurology, 14(3), 168–181. https://doi.org/10.1038/nrneurol.2017.187*
2. ***Huang, Y., & Mucke, L. (2012).*** *Alzheimer's mechanisms and therapeutic strategies. Cell, 148(6), 1204–1222. https://doi.org/10.1016/j.cell.2012.02.040*

3. **Long, J. M., & Holtzman, D. M. (2019).** *Alzheimer's disease: An update on pathobiology and treatment strategies. Cell, 179(2), 312–339. https://doi.org/10.1016/j.cell.2019.08.001*
4. **Palmqvist, S., Janelidze, S., Quiroz, Y. T., et al. (2020).** *Discriminative accuracy of plasma phospho-tau217 for Alzheimer's disease vs other neurodegenerative disorders. JAMA, 324(8), 772–781. https://doi.org/10.1001/jama.2020.10385*
5. **Rasmussen, K. L., Tybjærg-Hansen, A., Nordestgaard, B. G., & Frikke-Schmidt, R. (n.d.).** *APOE and dementia–resequencing and association studies in 65,000 individuals. Alzheimer's & Dementia. [Manuscript in preparation or submitted, no further information available.]*
6. **Selkoe, D. J., & Hardy, J. (2016).** *The amyloid hypothesis of Alzheimer's disease at 25 years. EMBO Molecular Medicine, 8(6), 595–608. https://doi.org/10.15252/emmm.201606210*

CHAPTER 11

Nutrition and Alzheimer's Disease

Welcome to the chapter on "Nutrition and Alzheimer's Disease," where we delve into the intricate relationship between diet and one of the most prevalent neurodegenerative conditions worldwide. Current and past research underscores the critical role of diet in shaping our body's biological processes, potentially influencing the development and progression of AD. Understanding how dietary choices interact with the body's mechanisms offers insights into strategies for potentially reducing or slowing down the AD symptoms through nutritional interventions.

In this Q&A chapter, we will explore the critical intersection of nutrition and Alzheimer's disease. We will examine how dietary factors influence the risk of developing Alzheimer's, explore the mechanisms through which specific nutrients affect brain health, and discuss dietary strategies that may support cognitive function.

Q1: How does diet influence Alzheimer's disease risk and progression?

Diet plays a crucial role in the risk and progression of Alzheimer's disease (AD). Numerous studies suggest that specific dietary patterns can either increase or decrease the risk of developing AD. Diets rich in fruits, vegetables, whole grains, lean proteins, and healthy fats, such as the Mediterranean and DASH (Dietary

Approaches to Stop Hypertension) diets, have been associated with a reduced risk of AD. These diets emphasize nutrient-dense foods that provide essential vitamins, minerals, antioxidants, and anti-inflammatory compounds, which are believed to protect against cognitive decline and neurodegeneration.

On the other hand, diets high in saturated fats, trans fats, refined sugars, and processed foods have been linked to an increased risk of AD. These foods can contribute to oxidative stress, inflammation, and vascular damage, which are key factors in AD pathogenesis. Emerging evidence also suggests that the gut– brain axis, which refers to the bidirectional communication between the gut and the brain, may play a role in AD. A healthy diet can promote a balanced gut microbiota, which may help protect against AD by reducing neuroinflammation and promoting brain health.

The standard Western diet, characterized by high consumption of red and processed meats, sugary snacks, and refined grains, is particularly concerning. This diet not only lacks essential nutrients but also promotes chronic diseases such as obesity, type 2 diabetes, and cardiovascular disease, all of which are risk factors for AD. The shift towards a Western diet in various cultures worldwide may contribute to the rising prevalence of AD, highlighting the need for dietary interventions that promote brain health.

▲ **FIGURE 11.1.**
Diet can affect Alzheimer's progress.

Q2: What are the key nutrients and dietary components that may help prevent Alzheimer's disease?

Several key nutrients and dietary components have been identified as potentially protective against Alzheimer's disease. These include Omega-3 fatty acids which are found in fatty fish, flaxseeds, and walnuts. Omega-3 fatty acids have anti-inflammatory properties and support brain health. Studies have shown that higher intake of omega-3s is associated with a lower risk of cognitive decline and AD. Omega-3 fatty acids, particularly docosahexaenoic acid (DHA), are integral components of neuronal membranes and are essential for synaptic function. They also help modulate the production of inflammatory cytokines, which are implicated in AD pathogenesis.

We also have antioxidants which can be found as vitamins E and C, which are found in nuts, seeds, fruits, and vegetables. They are potent antioxidants that can help reduce oxidative stress in the brain. Flavonoids, found in berries, tea, and dark chocolate, also have strong antioxidant effects and may protect against neurodegeneration. Antioxidants neutralize free radicals, which can cause cellular damage and contribute to the development of AD. The brain is particularly vulnerable to oxidative damage due to its high oxygen consumption and lipid-rich environment.

B vitamins, particularly B6, B12, and folate, are also essential for maintaining brain health and reducing homocysteine levels, which have been linked to an increased risk of AD. Leafy greens, legumes, and whole grains are good sources of these vitamins. Homocysteine is an amino acid that can damage blood vessels and neurons when present in high levels. B vitamins play a crucial role in converting homocysteine to methionine, thus, reducing its harmful effects.

Next, polyphenols are compounds which are found in a variety of plant foods such as fruits, vegetables, tea, coffee, and red wine, and have anti-inflammatory and neuroprotective effects. Resveratrol, a polyphenol found in red wine, has been shown to have potential benefits for brain health. Polyphenols can cross the blood–brain barrier and exert protective effects by reducing inflammation, improving blood flow, and modulating signaling pathways involved in neurodegeneration.

▲ **FIGURE 11.2.**
Balanced diets are beneficial.

Lastly, adequate levels of vitamin D, which can be obtained from sunlight exposure, fatty fish, and fortified foods, are important for brain health. Low levels of vitamin D have been associated with an increased risk of cognitive decline and AD. Vitamin D receptors are present in the brain, and the vitamin plays a role in neuroprotection, calcium homeostasis, and the regulation of immune responses.

Q3: How does the Mediterranean diet impact Alzheimer's disease risk?

The Mediterranean diet is widely recognized for its potential to reduce the risk of Alzheimer's disease. This diet emphasizes the consumption of fruits, vegetables, whole grains, legumes, nuts, seeds, olive oil, and moderate amounts of fish and poultry. Red meat and dairy products are consumed in smaller quantities, and red wine is consumed in moderation.

The Mediterranean diet's protective effects against AD are thought to result from its high content of antioxidants, anti-inflammatory compounds, and healthy fats. The diet provides ample amounts of vitamins C and E, polyphenols, and omega-3 fatty acids, which help reduce oxidative stress and inflammation in the brain. Additionally, the diet promotes cardiovascular health, which is closely linked to brain health, by reducing risk factors such as hypertension, high cholesterol, and obesity.

Several observational studies and clinical trials have shown that adherence to the Mediterranean diet is associated with a lower risk of cognitive decline and AD. For example, a study published in the *Journal of the American Medical Association* found that higher adherence to the Mediterranean diet was associated with a lower risk of AD and slower cognitive decline over time. Another study in the *American Journal of Clinical Nutrition* reported that individuals who closely followed the Mediterranean diet had a 33% lower risk of developing AD compared to those who did not adhere to the diet.

The Mediterranean diet also includes moderate alcohol consumption, particularly red wine, which contains resveratrol and other polyphenols that may protect against neurodegeneration. However, the benefits of moderate alcohol consumption must be balanced against the potential risks of alcohol abuse.

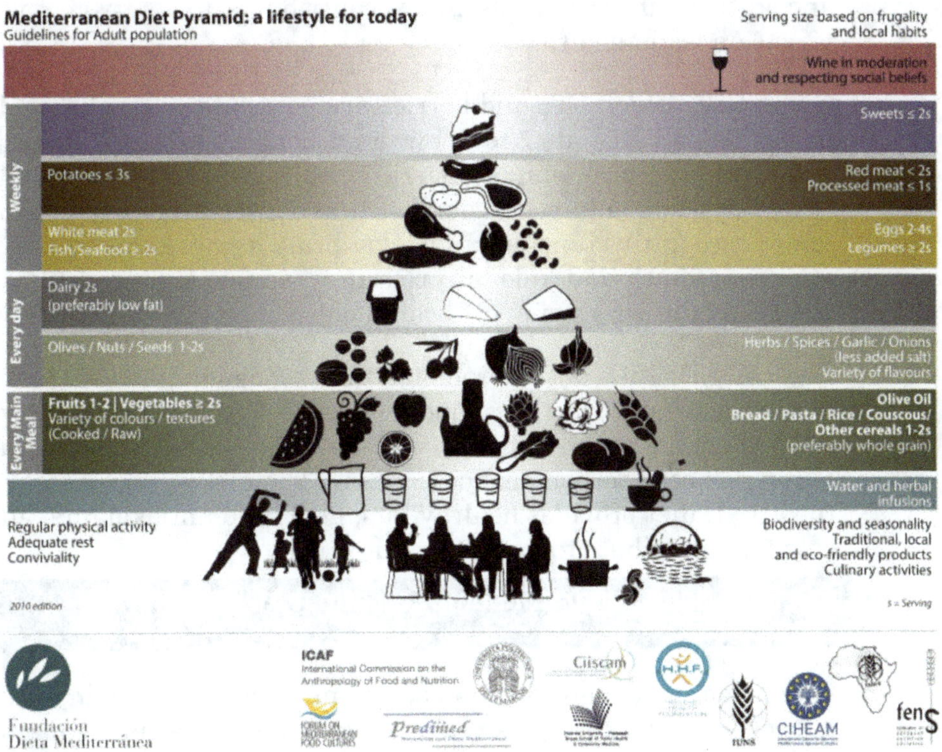

▲ **FIGURE 11.3.**
Mediterranean food pyramid.

Q4: Can the MIND diet help reduce Alzheimer's disease risk?

The MIND diet (Mediterranean-DASH Diet Intervention for Neurodegenerative Delay) is a hybrid of the Mediterranean and DASH diets, specifically designed to promote brain health and reduce the risk of Alzheimer's disease. The MIND diet emphasizes the consumption of ten brain-healthy food groups: green leafy vegetables, other vegetables, nuts, berries, beans, whole grains, fish, poultry, olive oil, and wine. It also recommends limiting the intake of red meat, butter and margarine, cheese, pastries and sweets, and fried or fast food.

Research on the MIND diet has shown promising results in reducing the risk of AD. A study published in *Alzheimer's & Dementia* found that high adherence to the MIND diet was associated with a 53% reduction in the risk of AD, while even moderate adherence was associated with a 35% reduction. The MIND diet's emphasis on nutrient-dense, anti-inflammatory, and antioxidant-rich foods likely contributes to its protective effects on brain health.

Green leafy vegetables and berries are particularly emphasized in the MIND diet due to their high content of antioxidants and bioactive compounds. Berries, such as blueberries and strawberries, contain flavonoids that have been shown to improve cognitive function and delay brain aging. Leafy greens, such as spinach and kale, are rich in vitamins A, C, and K, as well as folate, which support brain health.

The MIND diet also promotes the consumption of fish, particularly fatty fish like salmon, which is high in omega-3 fatty acids. Regular consumption of fish has been associated with a lower risk of cognitive decline and AD. The inclusion of nuts in the MIND diet provides healthy fats, fiber, and antioxidants, all of which contribute to overall brain health.

Q5: How does sugar consumption affect Alzheimer's disease risk?

High sugar consumption has been implicated in increasing the risk of Alzheimer's disease. Excessive intake of refined sugars and sugary beverages can lead to insulin resistance, obesity, type 2 diabetes, and metabolic syndrome, all of which are risk factors for AD. Insulin resistance, in particular, can impair brain function by reducing the effectiveness of insulin signaling in the brain, which is crucial for memory and cognitive function.

Moreover, high sugar consumption can contribute to oxidative stress and inflammation in the brain, promoting the formation of amyloid plaques and tau tangles, hallmark features of AD pathology. A study published in the *Journal of Alzheimer's Disease* found that higher consumption of sugary beverages was associated with poorer memory and smaller brain volume, suggesting a link between sugar intake and cognitive decline.

Reducing sugar intake and opting for a diet rich in whole, unprocessed foods can help mitigate these risks and support overall brain health. Replacing sugary snacks and beverages with fruits, vegetables, and whole grains can provide essential nutrients without the harmful effects of refined sugars. Additionally, managing blood sugar levels through a balanced diet can help reduce the risk of developing insulin resistance and type 2 diabetes, further protecting against AD.

Q6: What role does gut health play in Alzheimer's disease?

Emerging research suggests that gut health, particularly the composition and function of the gut microbiota, plays a significant role in Alzheimer's disease. The gut microbiota is a complex community of microorganisms residing in the gastrointestinal tract, and its balance can influence brain health through the gut–brain axis.

Dysbiosis, or an imbalance in the gut microbiota, has been associated with increased neuroinflammation, amyloid plaque formation, and cognitive decline. The gut microbiota can produce metabolites, such as short-chain fatty acids, that have anti-inflammatory and neuroprotective effects. Conversely, an unhealthy gut microbiota can produce harmful metabolites that contribute to inflammation and oxidative stress in the brain.

Diet plays a crucial role in shaping the gut microbiota. Diets rich in fiber, prebiotics, and probiotics, such as the Mediterranean diet, can promote a healthy gut microbiota and potentially protect against AD. A study published in the journal *Nutrients* found that adherence to the Mediterranean diet was associated with a more diverse and beneficial gut microbiota composition, which in turn was linked to better cognitive function.

Prebiotics, found in foods like garlic, onions, asparagus, and bananas, feed beneficial gut bacteria and promote their growth. Probiotics, found in fermented foods like yogurt, kefir, sauerkraut, and kimchi, introduce beneficial bacteria into the gut. Both prebiotics and probiotics can help maintain a balanced gut microbiota, supporting overall brain health and reducing the risk of AD.

Q7: How do dietary patterns impact Alzheimer's disease progression?

Dietary patterns, rather than individual nutrients, have a significant impact on Alzheimer's disease progression. Healthy dietary patterns, such as the Mediterranean and MIND diets, provide a combination of nutrients and bioactive compounds that work synergistically to support brain health. These diets emphasize whole, minimally processed foods that provide essential vitamins, minerals, antioxidants, and anti-inflammatory compounds.

Conversely, unhealthy dietary patterns, such as the Western diet, can accelerate AD progression by promoting oxidative stress, inflammation, and vascular damage. The Western diet is characterized by high consumption of processed foods, red and processed meats, sugary snacks, and refined grains, all of which can contribute to cognitive decline and neurodegeneration.

Adopting a healthy dietary pattern can help slow AD progression by providing the brain with the nutrients it needs to function optimally. For example, the Mediterranean diet's emphasis on fruits, vegetables, whole grains, and healthy fats provides a rich source of antioxidants and anti-inflammatory compounds that protect against neuronal damage. The MIND diet's focus on green leafy vegetables, berries, and nuts provides specific nutrients that support cognitive function and delay brain aging.

Q8: How does cultural diversity influence dietary practices and Alzheimer's disease risk?

Cultural diversity plays a significant role in dietary practices and, consequently, in the risk of Alzheimer's disease. Different cultures have unique dietary traditions and food preferences that can impact brain health. Understanding these cultural differences is essential for developing effective dietary interventions that are culturally relevant and acceptable.

Traditional diets in various cultures often include foods that promote brain health. For example, the traditional Japanese diet, which includes fish, rice, vegetables, and soy products, is rich in omega-3 fatty acids, antioxidants, and phytoestrogens, which have been linked to a lower risk of AD. Similarly, traditional Indian diets, which include a variety of spices, legumes, and whole grains, provide anti-inflammatory and neuroprotective compounds such as curcumin from turmeric.

In contrast, the adoption of Western dietary patterns, characterized by high consumption of processed foods, red meat, and sugary snacks, has been associated with an increased risk of AD in many cultures. This dietary shift can lead to an increase in chronic diseases such as obesity, type 2 diabetes, and cardiovascular disease, all of which are risk factors for AD.

To effectively reduce AD risk in diverse populations, it is important to promote dietary practices that align with cultural preferences and traditions. Public health initiatives should focus on educating individuals about the benefits of traditional diets and encouraging the incorporation of brain-healthy foods into daily meals. Additionally, culturally tailored dietary interventions can help address barriers to healthy eating and support long-term dietary changes.

Q9: What are the effects of intermittent fasting on Alzheimer's disease risk and progression?

Intermittent fasting, a dietary pattern that cycles between periods of fasting and eating, has gained attention for its potential neuroprotective effects. Research suggests that intermittent fasting may improve brain health by promoting autophagy (the body's way of cleaning out damaged cells), reducing inflammation, and enhancing mitochondrial function. Studies in animal models of Alzheimer's disease have shown that intermittent fasting can reduce amyloid beta accumulation and improve cognitive function. However, more human studies are needed to determine its efficacy and safety in AD prevention and management.

Q10: How do specific vitamins and minerals impact cognitive function and Alzheimer's disease risk?

Vitamins and minerals play crucial roles in brain health and cognitive function. For example, vitamin B12 and folate are essential for homocysteine metabolism, and deficiencies in these vitamins can lead to elevated homocysteine levels, which are associated with an increased risk of cognitive decline and

AD. Vitamin D is important for neuroprotection and anti-inflammatory responses. Minerals like magnesium and zinc are involved in synaptic plasticity and neurotransmission. Ensuring adequate intake of these vitamins and minerals through diet or supplementation may help reduce AD risk and support cognitive health.

Q11: How do polyphenols contribute to brain health and Alzheimer's disease prevention?

Polyphenols are bioactive compounds found in plant-based foods, such as fruits, vegetables, tea, coffee, and red wine. They have strong antioxidant and anti-inflammatory properties that can protect brain cells from oxidative stress and inflammation, both of which are implicated in AD. Specific polyphenols, like resveratrol (found in grapes and red wine) and epigallocatechin gallate (EGCG, found in green tea), have been studied for their neuroprotective effects. Research suggests that these compounds may help reduce amyloid beta accumulation, improve cognitive function, and lower the risk of AD.

Q12: What is the impact of caffeine and coffee consumption on Alzheimer's disease risk and progression?

Caffeine, found in coffee and other beverages, has been studied for its potential neuroprotective effects. Research suggests that caffeine may reduce amyloid beta aggregation, enhance cognitive function, and lower the risk of Alzheimer's disease. Understanding the relationship between caffeine consumption and brain health can provide insights into potential dietary strategies for Alzheimer's prevention.

Q13: How does protein intake affect Alzheimer's disease risk and progression?

Protein intake is essential for maintaining muscle mass, supporting immune function, and providing amino acids necessary for neurotransmitter synthesis. Adequate protein intake is particularly important in older adults to prevent sarcopenia (age-related muscle loss) and frailty, which can impact overall health and cognitive function. Some studies suggest that higher protein intake may be associated with a lower risk of cognitive decline. However, the type and source of protein (e.g., animal vs. plant-based) may also play a role in modulating AD risk and progression.

Q14: What is the impact of alcohol consumption on Alzheimer's disease risk?

Alcohol consumption has a complex relationship with brain health and AD risk. Moderate alcohol consumption, particularly of red wine, has been associated with a lower risk of AD and other forms of dementia, potentially due to the beneficial effects of polyphenols like resveratrol. However, heavy or excessive alcohol consumption is a significant risk factor for cognitive decline and dementia. Chronic alcohol abuse can lead to neuroinflammation, oxidative stress, and brain atrophy. The type and amount of alcohol consumed are critical factors in determining its impact on AD risk.

Q15: How does hydration status affect cognitive function and Alzheimer's disease risk?

Proper hydration is essential for maintaining cognitive function and overall brain health. Dehydration can impair attention, memory, and executive function, which are critical for daily activities. In older adults, dehydration is a common issue that can exacerbate cognitive decline and increase the risk of AD. Ensuring adequate fluid intake, primarily through water, is important for supporting brain function and reducing the risk of cognitive impairments.

Q16: What role do flavonoids play in brain health and Alzheimer's disease prevention?

Flavonoids are a diverse group of phytonutrients found in various fruits, vegetables, tea, and chocolate. They have potent antioxidant and anti-inflammatory properties that can protect neurons from damage. Research suggests that flavonoid-rich foods, such as berries, apples, and citrus fruits, are associated with a lower risk of cognitive decline and AD. Specific flavonoids, like quercetin and anthocyanins, have been shown to improve memory and cognitive function in animal models and human studies.

Q17: How do food additives and preservatives impact Alzheimer's disease risk?

Food additives and preservatives, commonly found in processed foods, have raised concerns about their potential impact on brain health. Some additives, such as artificial sweeteners and certain preservatives, may have neurotoxic effects and contribute to inflammation and oxidative stress. While more research is needed to fully understand the long-term effects of food additives on AD risk, minimizing the intake of highly processed foods and choosing whole, natural foods may support better brain health and reduce the risk of cognitive decline.

Q18: How does the ketogenic diet affect Alzheimer's disease risk and progression?

The ketogenic diet, which is high in fats and low in carbohydrates, has been studied for its potential benefits in neurological disorders, including AD. The diet induces a state of ketosis, where the body uses ketones as an alternative energy source to glucose. Ketones have been shown to provide neuroprotective effects, reduce amyloid beta levels, and improve mitochondrial function. Some studies suggest that the ketogenic diet may improve cognitive function in individuals with mild cognitive impairment and AD. However, more research is needed to establish its long-term safety and efficacy.

Q19: How do socioeconomic factors influence dietary habits and Alzheimer's disease risk?

Socioeconomic factors, such as income, education, and access to healthy foods, play a significant role in dietary habits and AD risk. Individuals with lower socioeconomic status may have limited access to nutritious foods, leading to poor dietary patterns that increase the risk of chronic diseases and cognitive decline. Education and income levels can also influence health literacy and the ability to make informed dietary choices. Addressing socioeconomic disparities and improving access to healthy foods are crucial for reducing AD risk and promoting brain health across diverse populations.

Q20: How do different types of fats (saturated, unsaturated, trans fats) impact Alzheimer's disease risk?

Different types of dietary fats have varying effects on brain health. Saturated and trans fats are associated with increased inflammation and higher risk of cognitive decline. In contrast, unsaturated fats, particularly omega-3 fatty acids found in fish, nuts, and seeds, are linked to improved brain function and reduced Alzheimer's disease risk. Understanding the impact of these fats can guide dietary choices to support cognitive health.

Q21: How does iron metabolism affect Alzheimer's disease risk and progression?

Iron is essential for brain function, but dysregulated iron metabolism can contribute to oxidative stress and neurodegeneration. Elevated brain iron levels have been observed in Alzheimer's disease and are thought to exacerbate amyloid beta aggregation and tau pathology. Maintaining proper iron balance through diet and, if necessary, medical intervention, can be important for managing Alzheimer's disease risk.

Q22: What are the future directions for research on nutrition and Alzheimer's disease?

Future research on nutrition and Alzheimer's disease will likely focus on several key areas including precision nutrition, gut–brain axis, nutrient synergy, longitudinal studies, mechanistic studies, intervention trials, and public health strategies.

Precision nutrition includes personalized dietary recommendations based on genetic, metabolic, and microbiome profiles that have the potential to optimize brain health and reduce AD risk. Precision nutrition approaches can help identify individual dietary needs and tailor interventions accordingly. For example, individuals with specific genetic variants that affect nutrient metabolism may benefit from targeted dietary strategies.

Gut–brain axis involves further exploration of the gut–brain axis and its role in AD to provide insights into how dietary interventions can modulate gut microbiota and influence brain health. Probiotic and prebiotic therapies may emerge as potential strategies for AD prevention and management. Understanding the complex interactions between the gut microbiota and the brain will be crucial for developing effective dietary interventions.

Nutrient synergy includes an approach investigating the synergistic effects of multiple nutrients and dietary patterns on brain health that will help identify the most effective combinations for reducing AD risk. Understanding how different nutrients interact and influence cognitive function is crucial for developing comprehensive dietary guidelines. For example, the combined effects of antioxidants, omega-3 fatty acids, and polyphenols may offer greater neuroprotection than any single nutrient alone.

Long-term studies better known as *longitudinal studies* are important to track dietary habits and cognitive health over time. This will provide valuable data on the impact of diet on AD risk and progression. These studies can help establish causal relationships and identify critical windows for dietary interventions. Longitudinal research will also help determine the long-term sustainability and effectiveness of specific dietary patterns.

Mechanistic studies involve research on the molecular mechanisms underlying the effects of diet on AD that have the potential to enhance our understanding of how specific nutrients and dietary components influence brain health. Identifying these mechanisms can inform the development of targeted nutritional therapies. For example, studying the effects of omega-3 fatty acids on neuronal function and inflammation may lead to novel therapeutic approaches.

Intervention trials are randomized controlled trials evaluating the efficacy of dietary interventions in preventing and managing AD that will provide robust evidence for dietary recommendations. These trials can potentially take a holistic approach and involve diverse populations and include long-term follow-up to assess the sustained impact of dietary changes. Well-designed intervention trials will help establish evidence-based dietary guidelines for AD prevention and management.

Public health strategies are also quite informative and important when it comes to future directions on AD research. This includes studying the impact of policy changes, educational programs, and community-based interventions on dietary behaviors and cognitive health. Public health initiatives should aim to make healthy eating accessible, affordable, and culturally acceptable for all populations.

In closing, the relationship between nutrition and Alzheimer's disease underscores the importance of a holistic approach to brain health. As we continue to uncover the intricate connections between diet, lifestyle, and neurodegeneration, it becomes evident that nutrition plays a pivotal role in both the prevention and management of Alzheimer's disease. By adopting healthy dietary patterns, including those rich in antioxidants, omega-3 fatty acids, and neuroprotective compounds, individuals may enhance brain resilience and potentially reduce their risk of cognitive decline.

This chapter has aimed to provide a comprehensive overview of how nutrition impacts Alzheimer's disease, from understanding dietary influences on disease pathology to practical recommendations for maintaining brain health. By integrating this knowledge into our everyday practices and keeping up with ongoing research efforts, we move closer to a future where effective prevention and treatment strategies for Alzheimer's disease are within reach.

Bibliography

1. **Ayton, S., Faux, N. G., & Bush, A. I. (2015).** *Ferritin levels in the cerebrospinal fluid predict Alzheimer's disease outcomes and are regulated by APOE. Nature Communications, 6, 6760. https://doi.org/10.1038/ncomms7760*
2. **Bazinet, R. P., & Layé, S. (2014).** *Polyunsaturated fatty acids and their metabolites in brain function and disease. Nature Reviews Neuroscience, 15(12), 771–785. https://doi.org/10.1038/nrn3820*
3. **Belaidi, A. A., & Bush, A. I. (2016).** *Iron neurochemistry in Alzheimer's disease and Parkinson's disease: Targets for therapeutics. Journal of Neurochemistry, 139(S1), 179–197. https://doi.org/10.1111/jnc.13425*
4. **Berendsen, A. A., Kang, J. H., Feskens, E. J., de Groot, L. C., Grodstein, F., & Stampfer, M. J. (2017).** *Association of DASH and Mediterranean dietary patterns with cognitive decline in older persons. Archives of Neurology, 69(10), 1246–1256. https://doi.org/10.1001/archneurol.2012.384*
5. **Dalile, B., Van Oudenhove, L., Vervliet, B., & Verbeke, K. (2019).** *The role of short-chain fatty acids in microbiota-gut-brain communication. Nature Reviews Gastroenterology & Hepatology, 16(8), 461–478. https://doi.org/10.1038/s41575-019-0100-x*
6. **Hu, X., Wang, T., & Jin, F. (2016).** *Alzheimer's disease and gut microbiota. Science China Life Sciences, 59(10), 1006–1023. https://doi.org/10.1007/s11427-016-5097-9*
7. **Morris, M. C., Evans, D. A., Bienias, J. L., Tangney, C. C., Bennett, D. A., Wilson, R. S., ... & Scherr, P. A. (2003).** *Dietary fats and the risk of incident Alzheimer's disease. Archives of Neurology, 60(2), 194–200. https://doi.org/10.1001/archneur.60.2.194*
8. **Morris, M. C., Tangney, C. C., Wang, Y., Sacks, F. M., Bennett, D. A., & Aggarwal, N. T. (2015).** *MIND diet associated with reduced incidence of Alzheimer's disease. Alzheimer's & Dementia, 11(9), 1007–1014. https://doi.org/10.1016/j.jalz.2015.04.002*
9. **Pase, M. P., Himali, J. J., Jacques, P. F., DeCarli, C., Satizabal, C. L., Aparicio, H. J., ... & Seshadri, S. (2017).** *Sugar- and artificially sweetened beverages and the risks of incident stroke and dementia: A prospective cohort study. Stroke, 48(5), 1139–1146. https://doi.org/10.1161/STROKEAHA.116.015572*
10. **Scarmeas, N., Stern, Y., Tang, M. X., Mayeux, R., & Luchsinger, J. A. (2006).** *Mediterranean diet and risk for Alzheimer's disease. Annals of Neurology, 59(6), 912–921. https://doi.org/10.1002/ana.20813*

11. ***Simpson, T. (2019).*** *Gut microbiota and Alzheimer's disease: The gut-brain connection. Alzheimer's Research & Therapy, 11(1), 15. https://doi.org/10.1186/s13195-019-0456-3*

12. ***Solfrizzi, V., Agosti, P., Lozupone, M., Custodero, C., Schilardi, A., Valiani, V., ... & Panza, F. (2018).*** *Nutritional interventions and cognitive-related outcomes in patients with late-life cognitive disorders: A systematic review. Journal of Alzheimer's Disease, 64(S1), S453–S484. https://doi.org/10.3233/JAD-179905*

CHAPTER 12 Exercise and Physical Activity

Welcome to the chapter on "Exercise and Physical Activity," where we're going to focus on the compelling connection between physical activity and Alzheimer's disease (AD). As Alzheimer's disease continues to affect millions worldwide, understanding the potential benefits of exercise in preventing and managing this condition becomes increasingly crucial. Research has consistently shown that regular physical activity can positively impact brain health, potentially reducing the risk of Alzheimer's and slowing its progression. This chapter explores how different types of exercise can influence cognitive function, the brain's structure, and overall well-being in individuals with Alzheimer's. By integrating exercise into daily routines, we can harness its potential to improve the quality of life for those living with Alzheimer's and offer hope for a healthier future.

Q1: How does regular physical exercise influence brain health and cognitive function in individuals at risk for Alzheimer's disease?

Regular physical exercise is crucial for maintaining brain health and cognitive function, especially in individuals at risk for Alzheimer's disease. Exercise promotes neuroplasticity, which is the brain's ability to form new connections and reorganize itself in response to learning and experience. This process is essential for memory formation and cognitive flexibility. Additionally, exercise stimulates the release of brain-derived neurotrophic factor (BDNF), a protein that supports the growth, survival, and differentiation of neurons in the brain. BDNF also plays a key role in synaptic plasticity, enhancing communication between neurons and thereby improving cognitive function. Furthermore, physical activity enhances cardiovascular health by improving blood flow to the brain, which ensures a steady supply of oxygen and nutrients necessary for optimal brain function. Exercise also reduces inflammation and oxidative stress, both of which are implicated in neurodegenerative processes like Alzheimer's disease.

Q2: What types of aerobic exercises are most effective in reducing the risk of Alzheimer's disease, and why?

Aerobic exercises such as brisk walking, jogging, swimming, and cycling are highly effective in reducing the risk of Alzheimer's disease. These exercises improve cardiovascular fitness, increasing the heart's ability to pump blood efficiently throughout the body, including the brain. Improved blood flow to the brain enhances oxygen delivery and nutrient supply, which are essential for maintaining neuronal health and function. Aerobic exercise also stimulates the production of BDNF, which promotes neurogenesis (the formation of new neurons), synaptic plasticity (the ability of synapses to strengthen or weaken over time), and neuronal survival. Studies suggest that regular participation in aerobic activities may help reduce the accumulation of beta-amyloid plaques and tau tangles in the brain, which are pathological hallmarks of Alzheimer's disease. Additionally, aerobic exercise improves glucose metabolism and insulin sensitivity, both of which play crucial roles in brain energy metabolism and cognitive function.

Q3: How does strength training impact brain health and cognitive function in older adults?

Strength training, also known as resistance training or weightlifting, provides unique benefits for brain health and cognitive function in older adults. It helps build and maintain muscle mass and strength, which is important for overall physical function and independence. Beyond its physical benefits, strength training also promotes neuroplasticity by stimulating the release of growth factors such as insulin-like growth factor 1 (IGF-1). These growth factors support the survival and growth of neurons and enhance synaptic plasticity, improving cognitive function and memory. Additionally, strength training can help reduce age-related loss of white matter in the brain, which is associated with cognitive decline. By improving overall physical health and functional capacity, strength training contributes to better brain health and resilience against neurodegenerative diseases like Alzheimer's.

Q4: What role do balance and coordination exercises play in reducing the risk of falls and cognitive decline in older adults with Alzheimer's?

Balance and coordination exercises are crucial for reducing the risk of falls and maintaining cognitive function in older adults. These exercises improve proprioception (awareness of body position in space), muscle strength, and coordination, which are essential for maintaining mobility and preventing falls. Falls in older adults can lead to injuries, reduced mobility, and increased risk of cognitive decline. By enhancing balance and coordination, these exercises help older adults maintain independence and reduce the fear of falling, thereby promoting physical activity and social engagement. Engaging in regular balance and coordination exercises also supports cognitive function by stimulating brain regions involved in motor control and spatial awareness, contributing to overall brain health and reducing the risk of cognitive decline.

Q5: What are the psychological benefits of exercise for individuals with Alzheimer's disease and their caregivers?

Exercise offers significant psychological benefits for individuals with Alzheimer's disease and their caregivers. For individuals with Alzheimer's, physical activity improves mood, reduces symptoms of depression and anxiety, and enhances overall well-being. It provides a sense of accomplishment and maintains a routine that can help reduce agitation and behavioral symptoms. Caregivers also benefit from exercise by reducing their stress levels, improving their mood, and enhancing their ability to cope with caregiving challenges. Exercise provides a valuable opportunity for social interaction and engagement for both individuals with Alzheimer's and their caregivers, fostering a supportive environment and improving quality of life for all involved.

Q6: How can community-based exercise programs contribute to the well-being of older adults and individuals with Alzheimer's disease?

Community-based exercise programs play a crucial role in promoting the well-being of older adults and individuals with Alzheimer's disease. These programs provide structured physical activity tailored to the needs and abilities of participants, promoting physical fitness, social interaction, and cognitive stimulation. Community-based programs offer a supportive environment where participants can engage in supervised exercise sessions, group activities, and recreational outings, fostering a sense of belonging and reducing social isolation. For individuals with Alzheimer's disease, these programs provide opportunities for cognitive stimulation through activities that challenge memory, problem-solving skills, and coordination. By promoting physical activity and social engagement, community-based exercise programs enhance overall well-being, independence, and quality of life for older adults and individuals living with Alzheimer's disease.

Q7: What are the potential challenges and barriers to maintaining regular physical activity in older adults with Alzheimer's disease?

Maintaining regular physical activity can be challenging for older adults with Alzheimer's disease due to various factors. Cognitive impairment may affect their ability to follow exercise instructions, remember routines, and safely participate in physical activities. Behavioral symptoms such as apathy, agitation, or resistance to change may also pose barriers to engaging in exercise programs. Caregiver support and supervision are often necessary to ensure safety and adherence to exercise routines. Environmental factors, such as accessibility to exercise facilities and transportation, can also impact participation in physical activity. Addressing these challenges requires personalized approaches that accommodate the individual's cognitive abilities, preferences, and physical limitations while promoting the benefits of regular exercise for brain health and overall well-being.

Q8: How do yoga and mind-body exercises benefit cognitive function and emotional well-being in individuals with Alzheimer's disease?

Yoga and mind-body exercises, such as tai chi and qigong, offer significant benefits for cognitive function and emotional well-being in individuals with Alzheimer's disease. These practices emphasize gentle physical movements, controlled breathing, and mindfulness, which together enhance physical health and mental clarity. Yoga and mind-body exercises improve flexibility, balance, and coordination, reducing the risk of falls and promoting physical well-being. The mindful aspect of these exercises helps reduce stress, anxiety, and depression by promoting relaxation and emotional regulation. Engaging in regular yoga or mind-body exercises can enhance cognitive function by improving attention, memory, and executive function through the integration of physical movement and mental focus. These practices also foster a sense of calm and well-being, contributing to overall quality of life for individuals with Alzheimer's disease.

Q9: What are the effects of high-intensity interval training (HIIT) on brain health and cognitive function in older adults with Alzheimer's?

High-intensity interval training (HIIT), which involves short bursts of intense exercise followed by periods of rest or low-intensity activity, has been shown to have positive effects on brain health and cognitive function in older adults. HIIT improves cardiovascular fitness and increases blood flow to the brain, enhancing oxygen and nutrient delivery critical for neuronal health. This type of training stimulates the production of BDNF, which supports neuroplasticity, cognitive function, and memory. HIIT also promotes the reduction of inflammation and oxidative stress, processes that contribute to neurodegeneration. Research indicates that older adults who engage in HIIT may experience improvements in executive function, memory, and processing speed. The time-efficient nature of HIIT makes it an attractive option for older adults seeking to maximize the benefits of exercise in a shorter period.

Q10: How do physical activities like dancing and rhythmic movement benefit individuals with Alzheimer's disease?

Physical activities such as dancing and rhythmic movement offer unique benefits for individuals with Alzheimer's disease. Dancing involves coordination, balance, and memory, as participants learn and recall dance steps, which stimulates multiple areas of the brain simultaneously. The rhythmic nature of dancing can improve motor function and enhance neuroplasticity. Dancing also provides social interaction and emotional expression, reducing feelings of isolation and boosting mood. Engaging in rhythmic movement activities can enhance cognitive function by promoting brain connectivity and improving memory and attention. For individuals with Alzheimer's, dancing offers a joyful and engaging way to combine physical exercise with cognitive stimulation and social interaction, contributing to overall well-being.

Q11: How does aquatic exercise impact cognitive function and physical health in older adults with Alzheimer's disease?

Aquatic exercise, performed in water, is particularly beneficial for older adults with Alzheimer's disease due to its low-impact nature and supportive environment. The buoyancy of water reduces stress on joints, making it easier for individuals with mobility issues to participate in physical activity. Aquatic exercise improves cardiovascular fitness, strength, and flexibility, contributing to overall physical health. The resistance of water provides a gentle form of strength training, enhancing muscle tone and endurance. Engaging in aquatic exercise also promotes cognitive function by improving blood flow to the brain and stimulating neuroplasticity. Additionally, the calming effect of water can reduce anxiety and agitation, promoting emotional well-being. For individuals with Alzheimer's, aquatic exercise offers a safe and enjoyable way to maintain physical health and cognitive function.

Q12: What role does social interaction during group exercise play in the cognitive and emotional health of older adults?

Social interaction during group exercise plays a vital role in the cognitive and emotional health of older adults. Participating in group exercise classes fosters a sense of community and belonging, which can alleviate feelings of loneliness and isolation. Social engagement provides cognitive stimulation through conversation, shared activities, and the need to follow group instructions. This interaction can enhance mood, reduce symptoms of depression and anxiety, and improve overall mental well-being. For older adults, the combination of physical activity and social interaction offers a holistic approach to maintaining cognitive function and emotional health. Group exercise programs also provide accountability and motivation, encouraging regular participation and the sustained benefits of physical activity.

Q13: How does outdoor physical activity, such as walking in nature, impact mental health and cognitive function in older adults?

Outdoor physical activity, such as walking in nature, has a profound impact on mental health and cognitive function in older adults. Exposure to natural environments provides a calming effect, reducing stress and anxiety. Walking in nature promotes physical fitness, improves cardiovascular health, and enhances blood flow to the brain. The sensory stimulation from natural surroundings can boost mood and improve cognitive function by engaging different areas of the brain involved in processing sensory information. Studies suggest that spending time in nature can improve attention, memory, and executive function. For older adults, outdoor physical activity offers a refreshing and enjoyable way to combine exercise with the mental health benefits of being in natural settings.

Q14: What are the long-term benefits of maintaining a regular exercise routine for brain health and cognitive function in older adults?

Maintaining a regular exercise routine offers significant long-term benefits for brain health and cognitive function in older adults. Regular physical activity supports neuroplasticity, enhances synaptic plasticity, and promotes the growth of new neurons, all of which are crucial for maintaining cognitive function. Exercise improves cardiovascular health, ensuring efficient blood flow to the brain and reducing the risk of vascular-related cognitive decline. It also helps regulate glucose metabolism and insulin sensitivity, both of which are vital for brain energy metabolism. Long-term exercise reduces inflammation and oxidative stress, processes linked to neurodegeneration. Consistently engaging in physical activity can delay the onset of cognitive decline, reduce the risk of Alzheimer's disease, and enhance overall quality of life by promoting physical, mental, and emotional well-being.

Q15: How do different intensities of physical activity (light, moderate, vigorous) impact cognitive function and brain health?

Different intensities of physical activity have varying impacts on cognitive function and brain health. Light physical activity, such as casual walking or stretching, provides basic health benefits, including improved circulation and reduced sedentary behavior. Moderate physical activity, such as brisk walking or gardening, enhances cardiovascular fitness, supports weight management, and stimulates neuroplasticity, improving cognitive function and memory. Vigorous physical activity, such as running or high-intensity interval training, significantly boosts cardiovascular health, increases BDNF levels, and promotes neurogenesis and synaptic plasticity. While all levels of physical activity contribute to brain health, combining different intensities can provide a comprehensive approach to maintaining cognitive function and overall well-being.

Q16: What are the recommended guidelines for physical activity to support brain health in older adults?

The recommended guidelines for physical activity to support brain health in older adults suggest at least 150 minutes of moderate-intensity aerobic exercise or 75 minutes of vigorous-intensity aerobic exercise per week. This can be achieved through activities like brisk walking, swimming, cycling, or dancing. Additionally, it is recommended to engage in muscle-strengthening activities involving major muscle groups on two or more days per week. Balance and flexibility exercises, such as yoga or tai chi, should also be incorporated to enhance physical stability and reduce the risk of falls. Following these guidelines helps maintain cardiovascular health, supports neuroplasticity, and promotes overall brain health and cognitive function in older adults.

Q17: How does physical exercise influence the release of lactic acid and malic acid, and what are the implications of this on Alzheimer's disease patients?

Physical exercise is widely acknowledged for its comprehensive health benefits, including improvements in cardiovascular fitness, mood enhancement, and cognitive function. An essential aspect of exercise is its impact on the body's metabolic processes, particularly the production of lactic acid and malic acid. These metabolites, often associated with muscle fatigue and soreness, play significant roles in energy metabolism and cellular signaling. Understanding how these substances affect brain health, especially in Alzheimer's disease patients, reveals important insights into the potential therapeutic effects of physical activity on neurodegenerative conditions.

During intense physical activity, when oxygen levels are insufficient for aerobic respiration, the body primarily relies on anaerobic metabolism. This process converts glucose into pyruvate, which is subsequently reduced to lactic acid. The accumulation of lactic acid in muscles is a well-known phenomenon, often linked to the sensation of fatigue and muscle soreness. However, lactic acid is not merely a waste product; it has significant physiological functions, particularly in the brain. Lactic acid can cross the blood–brain barrier and serve as an alternative energy source for neurons, supplementing glucose metabolism. This capability is especially crucial for Alzheimer's disease patients, whose brains often exhibit impaired glucose metabolism. Moreover, lactic acid has been shown to promote the expression of brain-derived neurotrophic factor (BDNF), a protein vital for neurogenesis (the creation of new neurons) and synaptic plasticity (the ability of synapses to strengthen or weaken over time, essential for learning and memory). Enhanced levels of BDNF are associated with improved cognitive function and memory, which are critical for Alzheimer's patients.

Malic acid is an intermediate in the Krebs cycle (also known as the citric acid cycle), a fundamental component of cellular respiration that occurs in the mitochondria. This cycle is essential for converting carbohydrates, fats, and proteins into carbon dioxide, water, and energy in the form of adenosine triphosphate (ATP). During exercise, the increased demand for ATP accelerates the Krebs cycle, leading to higher levels of malic acid. In Alzheimer's disease, mitochondrial dysfunction and altered energy metabolism are significant pathological features. By enhancing the Krebs cycle, malic acid helps improve ATP production, which is vital for maintaining neuronal function in a disease. Moreover, physical activity improves mood, reduces symptoms of depression and anxiety, and enhances overall well-being, contributing to better management of the disease. Although exercise cannot cure Alzheimer's, it plays a significant role in mitigating its progression and improving the lives of those affected.

Q19: What are the potential risks and safety considerations for older adults with Alzheimer's disease engaging in physical activity?

While physical activity offers numerous benefits, there are potential risks and safety considerations for older adults with Alzheimer's disease. Cognitive impairment may affect their ability to understand and follow exercise instructions, increasing the risk of accidents and injuries. Balance and coordination issues can lead to falls, so it is important to create a safe exercise environment with proper supervision. Physical activity programs should be tailored to the individual's abilities and limitations, avoiding overly strenuous activities that may cause fatigue or physical strain. Regular monitoring by caregivers or exercise professionals can help ensure safety and prevent adverse events. Encouraging low-impact and supervised activities can help older adults with Alzheimer's disease engage in physical exercise safely and effectively.

Q20: How can caregivers support and encourage physical activity in individuals with Alzheimer's disease?

Caregivers play a crucial role in supporting and encouraging physical activity in individuals with Alzheimer's disease. They can create a structured routine that incorporates regular exercise, making it a consistent part of the daily schedule. Caregivers should choose activities that match the individual's abilities and interests, ensuring that the exercise is enjoyable and engaging. Providing positive reinforcement and encouragement can motivate individuals to participate in physical activity. Caregivers can also join in the exercise sessions, making it a shared activity that fosters social interaction and bonding. Ensuring a safe environment, offering gentle guidance, and being attentive to the individual's needs and preferences are essential for promoting regular physical activity and its benefits for cognitive and physical health.

Q21: What type of exercise is most beneficial for preventing Alzheimer's disease?

Both aerobic exercise and strength training have been shown to be beneficial in preventing Alzheimer's disease. Aerobic exercises, such as walking, running, and swimming, improve cardiovascular health and increase blood flow to the brain, which can help maintain cognitive function. Strength training helps build muscle mass and strength, which can also support overall brain health. A combination of both types of exercise is recommended for optimal benefits.

Q22: Can exercise help slow the progression of Alzheimer's disease?

Exercise can help slow the progression of Alzheimer's disease by improving cognitive function and reducing the severity of symptoms. Physical activity has been shown to enhance neuroplasticity, which is the brain's ability to adapt and reorganize itself. This can help mitigate the effects of Alzheimer's disease and improve quality of life for those living with the condition.

Q23: How often should individuals exercise to reduce the risk of Alzheimer's disease?

The American Heart Association recommends at least 150 minutes of moderate-intensity aerobic exercise or 75 minutes of vigorous-intensity aerobic exercise per week, along with muscle-strengthening activities on two or more days per week. This level of physical activity has been associated with a lower risk of Alzheimer's disease and other forms of dementia.

Q24: What role does exercise play in overall brain health?

Exercise plays a crucial role in overall brain health by promoting neurogenesis, improving blood flow, and reducing inflammation. Regular physical activity has been shown to enhance cognitive function, improve memory, and reduce the risk of cognitive decline. Additionally, exercise can help reduce stress and improve mood, which can further support brain health and overall well-being.

Q25: How does exercise influence the brain's structure and function in the context of Alzheimer's disease?

Exercise has a profound impact on both the structure and function of the brain, particularly in the context of Alzheimer's disease. Regular physical activity promotes neuroplasticity, the brain's ability to reorganize itself by forming new neural connections. This process is crucial for learning and memory, which are often impaired in Alzheimer's disease. Exercise also stimulates the production of brain-derived neurotrophic factor (BDNF), a protein that supports the survival of existing neurons and encourages the growth of new neurons and synapses. Furthermore, exercise improves cerebral blood flow, ensuring that the brain receives adequate oxygen and nutrients, which can help mitigate the cognitive decline associated with Alzheimer's. Studies using neuroimaging techniques have shown that individuals who engage in regular physical activity tend to have larger hippocampal volumes and greater overall brain volume, both of which are associated with better cognitive function.

Q26: What are the psychological benefits of exercise for individuals with Alzheimer's disease?

Exercise offers numerous psychological benefits for individuals with Alzheimer's disease, extending beyond the physical improvements. Engaging in regular physical activity can help alleviate symptoms of depression and anxiety, which are common comorbidities in Alzheimer's patients. Physical exercise triggers the release of endorphins, chemicals in the brain that act as natural painkillers and mood elevators. This can lead to improved mood and a greater sense of well-being. Additionally, exercise can enhance sleep quality, which is often disrupted in individuals with Alzheimer's disease. Better sleep contributes to improved cognitive function and overall mental health. Participating in group exercise activities also provides social interaction, reducing feelings of isolation and loneliness, and fostering a sense of community and support among participants.

Q27: How can caregivers incorporate exercise into the daily routine of individuals with Alzheimer's disease?

Caregivers play a crucial role in incorporating exercise into the daily routine of individuals with Alzheimer's disease. To start, it is important to choose activities that the individual enjoys and finds motivating. Simple exercises such as walking, dancing, or gardening can be effective and enjoyable. Establishing a regular schedule can help make exercise a consistent part of the daily routine. Caregivers can also look for opportunities to incorporate physical activity into daily tasks, such as encouraging the individual to walk around the house or participate in light household chores. Additionally, group exercise classes specifically designed for individuals with dementia can provide structured and safe environments for physical activity, while also offering social interaction. Caregivers should ensure that the exercise activities are safe and appropriate for the individual's physical abilities, and they should consult healthcare professionals for personalized recommendations.

Q28: What role does exercise play in managing behavioral symptoms of Alzheimer's disease?

Exercise plays a significant role in managing behavioral symptoms of Alzheimer's disease, such as agitation, aggression, and wandering. Physical activity can help reduce these symptoms by providing a constructive outlet for excess energy and frustration. Engaging in regular exercise can also help regulate sleep patterns, which in turn can reduce nighttime agitation and wandering. Moreover, exercise has been shown to have a calming effect, reducing anxiety and restlessness. Activities that involve rhythmic movement, such as walking or swimming, can be particularly soothing and help alleviate agitation. By improving overall physical health, exercise can also reduce pain and discomfort, which are often underlying causes of behavioral symptoms. Caregivers can use exercise as a tool to help manage these challenging behaviors, improving the quality of life for both the individual with Alzheimer's and their caregivers.

Q29: How does exercise impact inflammation and oxidative stress in the brain, and what implications does this have for Alzheimer's disease?

Exercise has been shown to reduce inflammation and oxidative stress in the brain, both of which are implicated in the pathogenesis of Alzheimer's disease. Physical activity stimulates the production of antioxidants, which help neutralize free radicals and reduce oxidative damage to brain cells. This protective effect can help preserve neuronal function and slow the progression of cognitive decline. Exercise also has anti-inflammatory effects, reducing the levels of pro-inflammatory cytokines in the brain. Chronic inflammation is a hallmark of Alzheimer's disease and contributes to neuronal damage and death. By reducing inflammation, exercise can help protect the brain from further damage and support overall cognitive health. These neuroprotective effects of exercise highlight its potential as a non-pharmacological intervention to mitigate the risk and progression of Alzheimer's disease.

In conclusion, exercise and physical activity play a vital role in promoting brain health and potentially mitigating the effects of Alzheimer's disease. From improving cognitive function and reducing behavioral symptoms to enhancing psychological well-being and promoting social interaction, the benefits of regular physical activity are profound. As we continue to uncover the mechanisms by which exercise influences brain health, it becomes increasingly clear that incorporating physical activity into our daily lives is a powerful tool in the fight against Alzheimer's. By prioritizing exercise and making it a fundamental part of our routines, we can support brain health, enhance quality of life, and contribute to a brighter future for those affected by Alzheimer's disease.

Bibliography

1. ***Ahlskog, J. E., Geda, Y. E., Graff-Radford, N. R., & Petersen, R. C. (2011).*** *Physical exercise as a preventive or disease-modifying treatment of dementia and brain aging. Mayo Clinic Proceedings, 86(9), 876–884. https://doi.org/10.4065/mcp.2011.0252*
2. ***Baker, L. D., Frank, L. L., Foster-Schubert, K., et al. (2010).*** *Effects of aerobic exercise on mild cognitive impairment: A controlled trial. Archives of Neurology, 67(1), 71–79. https://doi.org/10.1001/archneurol.2009.307*
3. ***Blondell, S. J., Hammersley-Mather, R., & Veerman, J. L. (2014).*** *Does physical activity prevent cognitive decline and dementia? A systematic review and meta-analysis of longitudinal studies. BMC Public Health, 14, 510. https://doi.org/10.1186/1471-2458-14-510*
4. ***Cassilhas, R. C., Lee, K. S., Fernandes, J., et al. (2007).*** *Spatial memory is improved by aerobic and resistance exercise through divergent molecular mechanisms. Neuroscience, 144(3), 1033–1042. https://doi.org/10.1016/j.neuroscience.2006.10.021*
5. ***Colcombe, S., & Kramer, A. F. (2003).*** *Fitness effects on the cognitive function of older adults: A meta-analytic study. Psychological Science, 14(2), 125–130. https://doi.org/10.1111/1467-9280.t01-1-01430*
6. ***Erickson, K. I., Voss, M. W., Prakash, R. S., et al. (2011).*** *Exercise training increases size of hippocampus and improves memory. Proceedings of the National Academy of Sciences, 108(7), 3017–3022. https://doi.org/10.1073/pnas.1015956108*
7. ***Geda, Y. E., Roberts, R. O., Knopman, D. S., et al. (2010).*** *Physical exercise, aging, and mild cognitive impairment: A population-based study. Archives of Neurology, 67(1), 80–86. https://doi.org/10.1001/archneurol.2009.307*
8. ***Guiney, H., & Machado, L. (2013).*** *Benefits of regular aerobic exercise for executive functioning in healthy populations. Psychonomic Bulletin & Review, 20(1), 73–86. https://doi.org/10.3758/s13423-012-0333-y*
9. ***Hamer, M., & Chida, Y. (2009).*** *Physical activity and risk of neurodegenerative disease: A systematic review of prospective evidence. Psychological Medicine, 39(1), 3–11. https://doi.org/10.1017/S0033291708003681*
10. ***Heyn, P., Abreu, B. C., & Ottenbacher, K. J. (2004).*** *The effects of exercise training on elderly persons with cognitive impairment and dementia: A meta-analysis. Archives of Physical Medicine and Rehabilitation, 85(10), 1694–1704. https://doi.org/10.1016/j.apmr.2003.12.020*
11. ***Larson, E. B., Wang, L., Bowen, J. D., et al. (2006).*** *Exercise is associated with reduced risk for incident dementia among persons 65 years of age and older. Annals of Internal Medicine, 144(2), 73–81. https://doi.org/10.7326/0003-4819-144-2-200601170-00004*
12. ***Lautenschlager, N. T., Cox, K. L., Flicker, L., et al. (2008).*** *Effect of physical activity on cognitive function in older adults at risk for Alzheimer's disease: A randomized trial. JAMA, 300(9), 1027–1037. https://doi.org/10.1001/jama.300.9.1027*
13. ***Middleton, L. E., & Yaffe, K. (2009).*** *Promising strategies for the prevention of dementia. Archives of Neurology, 66(10), 1210–1215. https://doi.org/10.1001/archneurol.2009.205*

14. ***Northey, J. M., Cherbuin, N., Pumpa, K. L., et al. (2018).*** *Exercise interventions for cognitive function in adults older than 50: A systematic review with meta-analysis. British Journal of Sports Medicine, 52(3), 154–160. https://doi.org/10.1136/bjsports-2017-098682*
15. ***Rolland, Y., Abellan van Kan, G., & Vellas, B. (2008).*** *Physical activity and Alzheimer's disease: From prevention to therapeutic perspectives. Journal of the American Medical Directors Association, 9(6), 390–405. https://doi.org/10.1016/j.jamda.2008.03.003*
16. ***Scherder, E., Eggermont, L., Swaab, D., et al. (2014).*** *Gait in ageing and associated dementias; its relationship with cognition. Neuroscience and Biobehavioral Reviews, 38, 1–16. https://doi.org/10.1016/j.neubiorev.2013.12.003*
17. ***Sink, K. M., Espeland, M. A., Castro, C. M., et al. (2015).*** *Effect of a 24-month physical activity intervention vs health education on cognitive outcomes in sedentary older adults: The LIFE randomized trial. JAMA, 314(8), 781–790. https://doi.org/10.1001/jama.2015.9616*
18. ***Smith, P. J., Blumenthal, J. A., Hoffman, B. M., et al. (2010).*** *Aerobic exercise and neurocognitive performance: A meta-analytic review of randomized controlled trials. Psychosomatic Medicine, 72(3), 239–252. https://doi.org/10.1097/PSY.0b013e3181d1461e*
19. ***Tanaka, H., & Seals, D. R. (2008).*** *Endurance exercise performance in Masters athletes: Age-associated changes and underlying physiological mechanisms. Journal of Physiology, 586(1), 55–63. https://doi.org/10.1113/jphysiol.2007.140208*
20. ***Tarumi, T., & Zhang, R. (2014).*** *The role of exercise-induced cardiovascular adaptation in brain health. Exercise and Sport Sciences Reviews, 42(4), 172–179. https://doi.org/10.1249/JES.0000000000000030*

CHAPTER 13 *Prevention Strategies and Risk Reduction*

As the global population ages, the prevalence of Alzheimer's is expected to increase significantly, posing substantial challenges to healthcare systems and caregivers. Given the lack of a definitive cure, the emphasis has shifted towards prevention and risk reduction strategies that can delay the onset and progression of the disease. In this chapter we will be exploring "Prevention Strategies and Risk Reduction" mechanisms. Research has identified a range of modifiable risk factors and lifestyle changes that can significantly impact an individual's likelihood of developing Alzheimer's. These include regular physical exercise, maintaining a balanced diet, engaging in cognitive training exercises, fostering social connections, ensuring adequate sleep, and managing stress. Each of these factors contributes to overall brain health by enhancing neuroplasticity, reducing inflammation, and supporting cardiovascular health. Understanding and implementing these prevention strategies can empower individuals to take proactive steps towards protecting their cognitive health. Moreover, public health initiatives and educational campaigns can play a crucial role in raising awareness about Alzheimer's prevention, encouraging healthier lifestyles, and providing resources and support for at-risk populations. This holistic approach not only benefits individuals but also has the potential to reduce the societal and economic burden associated with Alzheimer's disease. As research continues to unravel the complex interplay between genetic, environmental, and lifestyle factors in Alzheimer's development, it becomes increasingly clear that a multifaceted approach is necessary. By addressing the various risk factors through comprehensive prevention strategies, we can move closer to mitigating the impact of Alzheimer's and enhancing the quality of life for individuals as they age.

Q1: What are the primary lifestyle changes recommended to reduce the risk of Alzheimer's disease?

The primary lifestyle changes recommended to reduce the risk of Alzheimer's disease focus on maintaining overall health and brain function. Regular physical exercise is crucial, as it improves cardiovascular health, increases blood flow to the brain, and enhances brain plasticity. A balanced diet, such as the Mediterranean diet, which is rich in fruits, vegetables, whole grains, lean proteins, and healthy fats, can reduce inflammation and provide essential nutrients for brain health. Cognitive engagement through activities like puzzles, reading, and learning new skills can stimulate the brain and maintain cognitive function. Social interaction is equally important, as it helps reduce stress and depression, both of which are risk factors for Alzheimer's. Adequate sleep is essential, as poor sleep can lead to the accumulation of amyloid-beta plaques, a hallmark of Alzheimer's. Finally, stress management techniques such as mindfulness, meditation, and regular relaxation can lower the risk by reducing chronic stress and its negative effects on the brain.

Q2: How does physical exercise help in reducing the risk of Alzheimer's?

Physical exercise is one of the most effective strategies for reducing the risk of Alzheimer's disease. Regular physical activity promotes cardiovascular health, which is closely linked to brain health. Improved blood flow resulting from exercise ensures that the brain receives sufficient oxygen and nutrients, which are crucial for maintaining cognitive function. Exercise also stimulates the production of neurotrophic factors, such as brain-derived neurotrophic factor (BDNF), which support the growth and survival of neurons. Additionally, physical activity helps to reduce inflammation and oxidative stress, both of which are implicated in Alzheimer's pathology. Exercise can also enhance brain plasticity, the brain's ability to adapt and reorganize itself, which is essential for learning and memory. Moreover, regular physical activity can help manage other risk factors for Alzheimer's, such as obesity, diabetes, and hypertension, by maintaining a healthy weight and improving metabolic health. Overall, the combination of direct benefits to brain health and the management of comorbid conditions makes physical exercise a powerful tool in reducing the risk of Alzheimer's disease.

Q3: What role does diet play in Alzheimer's prevention?

Diet plays a significant role in Alzheimer's prevention by influencing brain health through various mechanisms. Diets rich in fruits, vegetables, whole grains, lean proteins, and healthy fats provide essential nutrients that support brain function and reduce inflammation. Antioxidants found in fruits and vegetables help combat oxidative stress, which can damage brain cells and contribute to Alzheimer's disease. Omega-3 fatty acids, commonly found in fish, nuts, and seeds, are crucial for maintaining neuronal membrane integrity and promoting anti-inflammatory pathways in the brain. The Mediterranean diet, in particular, has been associated with a reduced risk of Alzheimer's due to its emphasis on these nutritious foods and its ability to improve cardiovascular health. A healthy diet also helps manage other risk factors for Alzheimer's, such as hypertension, diabetes, and obesity. For example, reducing sugar intake can help regulate blood glucose levels and prevent insulin resistance, which is linked to cognitive decline. By maintaining overall metabolic health and providing the brain with necessary nutrients, a well-balanced diet serves as a key component in Alzheimer's prevention strategies.

Q4: Can cognitive training exercises reduce Alzheimer's risk?

Yes, cognitive training exercises can reduce the risk of Alzheimer's disease by keeping the brain engaged and improving cognitive reserve. Cognitive reserve refers to the brain's ability to compensate for age-related changes and neurodegenerative processes. Engaging in mentally stimulating activities such as puzzles, memory games, learning new skills, or even taking up new hobbies can enhance neural connections and promote brain plasticity. These activities stimulate various cognitive functions, including memory, attention, language, and problem-solving skills, which can help delay the onset of Alzheimer's symptoms. Additionally, cognitive training exercises can promote the growth of new neurons and synapses, particularly in brain regions associated with memory and executive function. Regular mental stimulation also encourages the brain to develop alternative pathways for processing information, providing a buffer against cognitive decline. Research has shown that individuals who consistently engage in cognitive activities throughout their lives tend to have a lower risk of developing Alzheimer's disease. Therefore, incorporating cognitive training exercises into daily routines is an effective strategy for maintaining brain health and reducing the risk of Alzheimer's.

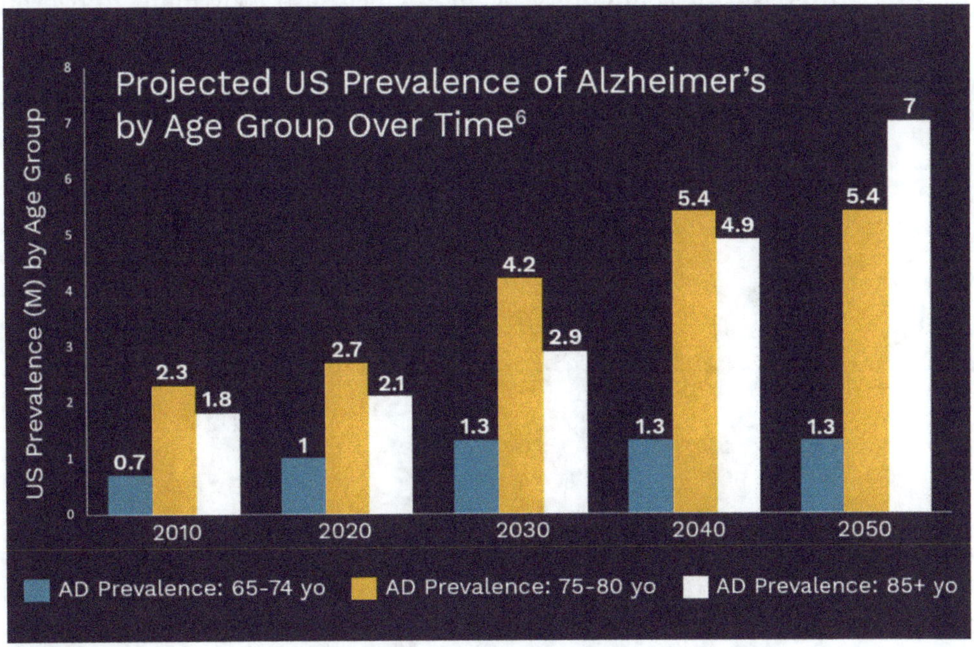

▲ **FIGURE 13.1.**
The risk of developing Alzheimer's disease as age increases.

Q5: How does social interaction influence the risk of Alzheimer's?

Social interaction plays a crucial role in reducing the risk of Alzheimer's disease by providing mental stimulation and emotional support. Engaging in regular social activities helps keep the brain active and challenged, which can enhance cognitive reserve and delay the onset of Alzheimer's symptoms. Social interactions involve communication, problem-solving, and emotional regulation, all of which stimulate different areas of the brain and promote neural connectivity. Additionally, maintaining strong social connections can reduce stress and depression, both of which are risk factors for Alzheimer's. Chronic stress and depression can lead to elevated levels of cortisol, a stress hormone that can damage brain cells and impair memory and cognitive function. By reducing stress and providing emotional support, social interactions can help protect the brain from these negative effects. Furthermore, social engagement often involves physical activities, such as group exercises or community events, which also contribute to overall brain health. Studies have shown that individuals who have strong social networks and engage in regular social activities have a lower risk of developing Alzheimer's disease. Therefore, fostering social connections and participating in community activities are important strategies for Alzheimer's prevention.

▲ **FIGURE 13.2.**
Displaying the purple ribbon showing support against Alzheimer's disease, showing hope for the future.

Q6: What is the impact of sleep on Alzheimer's risk?

Sleep has a profound impact on Alzheimer's risk, as it is crucial for brain health and the prevention of neurodegenerative diseases. During sleep, the brain undergoes important processes such as the clearance of metabolic waste products, including amyloid-beta, a protein that accumulates in Alzheimer's disease. Poor sleep quality or insufficient sleep can lead to the accumulation of amyloid-beta plaques, which are a hallmark of Alzheimer's. Additionally, sleep is essential for memory consolidation and cognitive function. Disrupted sleep patterns, such as those caused by sleep apnea or insomnia, can impair these processes and increase the risk of cognitive decline. Chronic sleep deprivation can also elevate stress hormone levels, leading to inflammation and oxidative stress, both of which are associated with Alzheimer's pathology. Establishing good sleep hygiene, such as maintaining a regular sleep schedule, creating a restful sleep environment, and avoiding stimulants before bedtime, can significantly improve sleep quality. By ensuring adequate and restorative sleep, individuals can reduce their risk of developing Alzheimer's disease and support overall brain health.

Q7: How does stress management contribute to reducing Alzheimer's risk?

Stress management is a crucial component of reducing the risk of Alzheimer's disease, as chronic stress can have detrimental effects on brain health. Prolonged stress leads to elevated levels of cortisol, a hormone that, in high amounts, can damage the hippocampus, a brain region essential for memory and learning. Chronic stress is also associated with increased inflammation and oxidative stress, which are risk factors for Alzheimer's. Effective stress management techniques, such as mindfulness, meditation, deep breathing exercises, and physical activities, can help lower cortisol levels and reduce the impact of stress on the brain. Additionally, engaging in hobbies, spending time in nature, and maintaining social connections can provide emotional support and reduce stress. By managing stress effectively, individuals can protect their brain from the harmful effects of chronic stress and lower their risk of Alzheimer's disease. Incorporating stress-reducing practices into daily routines is essential for maintaining mental and emotional well-being, which in turn supports cognitive health and resilience against neurodegenerative diseases.

Q8: What is the role of cardiovascular health in Alzheimer's prevention?

Cardiovascular health plays a significant role in Alzheimer's prevention because the brain relies on a healthy vascular system to receive adequate oxygen and nutrients. Conditions such as hypertension, heart disease, stroke, and diabetes can impair blood flow to the brain and increase the risk of cognitive decline. Maintaining cardiovascular health through regular physical activity, a balanced diet, and proper management of blood pressure, cholesterol, and blood sugar levels can help protect the brain from these risks. Exercise, in particular, improves heart health and enhances blood circulation, ensuring that the brain gets sufficient oxygen and nutrients. A heart-healthy diet, such as the Mediterranean diet, which is rich in fruits, vegetables, whole grains, and healthy fats, can also reduce inflammation and support vascular health. By promoting cardiovascular health, individuals can reduce their risk of developing Alzheimer's disease and support overall brain function. Regular check ups and monitoring of cardiovascular risk factors are essential for early detection and management, further contributing to the prevention of Alzheimer's disease.

Q9: How does managing diabetes affect Alzheimer's risk?

Managing diabetes is crucial for reducing Alzheimer's risk because diabetes is associated with an increased risk of cognitive decline and dementia. High blood sugar levels can lead to insulin resistance, which affects the brain's ability to use glucose effectively. This can result in impaired cognitive function and an increased risk of developing Alzheimer's disease. Additionally, diabetes-related complications such as cardiovascular disease and inflammation can further contribute to the risk. Effective management of diabetes through medication, a healthy diet, regular physical activity, and monitoring blood sugar levels can help prevent these complications. A diet low in refined sugars and rich in whole foods can improve insulin sensitivity and support overall metabolic health. Regular exercise enhances insulin sensitivity and helps regulate blood sugar levels. By managing diabetes effectively, individuals can reduce the risk of cognitive decline and protect their brain health, thereby lowering the risk of Alzheimer's disease.

Q10: What are the benefits of the Mediterranean diet in Alzheimer's prevention?

The Mediterranean diet has been shown to have numerous benefits in Alzheimer's prevention due to its emphasis on whole foods and healthy fats. This diet is rich in fruits, vegetables, whole grains, legumes, nuts, seeds, fish, and olive oil, and includes moderate consumption of dairy and red wine. The high intake of antioxidants from fruits and vegetables helps combat oxidative stress, which is implicated in Alzheimer's pathology. Omega-3 fatty acids, commonly found in fish and olive oil, have anti-inflammatory properties and support neuronal health. The Mediterranean diet also promotes cardiovascular health by reducing the risk of hypertension, heart disease, and stroke, all of which are risk factors for Alzheimer's. Additionally, this diet helps maintain a healthy weight and regulate blood sugar levels, further reducing the risk of diabetes and metabolic syndrome. Studies have shown that adherence to the Mediterranean diet is associated with a lower risk of cognitive decline and Alzheimer's disease. By incorporating the principles of this diet into daily eating habits, individuals can support brain health and reduce their risk of developing Alzheimer's.

Q11: How does hypertension impact Alzheimer's risk, and what can be done to manage it?

Hypertension, or high blood pressure, is a significant risk factor for Alzheimer's disease because it can damage blood vessels in the brain, leading to reduced blood flow and increased risk of stroke and other cerebrovascular conditions. Over time, this vascular damage can contribute to cognitive decline and the development of Alzheimer's. Managing hypertension is crucial for reducing these risks. This can be achieved through lifestyle modifications such as a balanced diet low in salt, regular physical exercise, maintaining a healthy weight, reducing alcohol consumption, and avoiding smoking. Medications prescribed by healthcare providers can also help manage blood pressure effectively. Regular monitoring of blood pressure levels and adherence to treatment plans are essential. By managing hypertension, individuals can protect their brain from vascular damage, reduce the risk of cognitive decline, and lower the likelihood of developing Alzheimer's disease.

Q12: What role does smoking cessation play in Alzheimer's prevention?

Smoking cessation is a critical factor in Alzheimer's prevention because smoking has been linked to an increased risk of cognitive decline and dementia. Smoking can lead to vascular damage, reducing blood flow to the brain and increasing the risk of stroke and other cerebrovascular conditions. It also promotes oxidative stress and inflammation, both of which are associated with Alzheimer's pathology. Quitting smoking can significantly improve cardiovascular health, reduce inflammation, and enhance overall brain health. The benefits of smoking cessation are evident at any age, but the earlier one quits, the better the outcomes for brain health. Additionally, smoking cessation can improve lung function, reduce the risk of chronic diseases, and enhance overall well-being. Support from healthcare providers, smoking cessation programs, and lifestyle changes can help individuals quit smoking and reduce their risk of developing Alzheimer's disease.

Q13: How can maintaining a healthy weight reduce Alzheimer's risk?

Maintaining a healthy weight is important for reducing Alzheimer's risk because obesity is associated with several conditions that increase the risk of cognitive decline and dementia, including diabetes, hypertension, and cardiovascular disease. Excess body weight can lead to insulin resistance, inflammation, and metabolic syndrome, all of which negatively impact brain health. Adopting a healthy diet, engaging in regular physical activity, and making lifestyle changes to achieve and maintain a healthy weight can improve metabolic health and reduce these risks. A balanced diet rich in whole foods, such as the Mediterranean diet, and regular exercise can help regulate blood sugar levels, reduce inflammation, and improve cardiovascular health. By maintaining a healthy weight, individuals can protect their brain from the adverse effects of obesity-related conditions and lower their risk of developing Alzheimer's disease.

Q14: What is the significance of monitoring cholesterol levels in Alzheimer's prevention?

Monitoring cholesterol levels is significant in Alzheimer's prevention because high cholesterol levels can contribute to vascular damage and increase the risk of stroke and other cerebrovascular conditions. Elevated cholesterol can lead to the buildup of plaques in blood vessels, reducing blood flow to the brain and increasing the risk of cognitive decline. By maintaining healthy cholesterol levels through a balanced diet, regular physical activity, and, if necessary, medications prescribed by healthcare providers, individuals can reduce these risks. A diet low in saturated fats and trans fats and rich in fruits, vegetables, whole grains, and healthy fats can help manage cholesterol levels effectively. Regular check ups and monitoring of cholesterol levels are essential for early detection and management. By maintaining healthy cholesterol levels, individuals can protect their brain from vascular damage and reduce the risk of Alzheimer's disease.

Q15: How does alcohol consumption affect Alzheimer's risk?

Alcohol consumption can affect Alzheimer's risk in different ways, depending on the amount and frequency of intake. Moderate alcohol consumption, particularly of red wine, has been associated with a reduced risk of Alzheimer's due to the presence of antioxidants like resveratrol, which may have neuroprotective effects. However, excessive alcohol consumption is a significant risk factor for cognitive decline and dementia. Heavy drinking can lead to brain damage, including the loss of brain cells and reduced brain volume, and it can impair cognitive function. Additionally, excessive alcohol consumption can increase the risk of cardiovascular conditions and liver disease, both of which are associated with an increased risk of Alzheimer 's. To reduce the risk, it is important to consume alcohol in moderation, following recommended guidelines, or to avoid it altogether if there are existing health concerns or a history of alcohol abuse.

Q16: What is the role of mental health in Alzheimer's prevention?

Mental health plays a crucial role in Alzheimer's prevention because conditions such as depression, anxiety, and chronic stress are associated with an increased risk of cognitive decline and dementia. Poor mental health can lead to elevated levels of stress hormones, such as cortisol, which can damage brain cells and impair cognitive function. Additionally, mental health conditions can lead to social isolation, reduced physical activity, and poor sleep, all of which are risk factors for Alzheimer's. Managing mental health through therapy, medication, stress management techniques, regular physical activity, and social engagement can help reduce these risks. Mindfulness, meditation, and other relaxation techniques can also improve mental well-being and reduce stress. By prioritizing mental health and seeking appropriate treatment and support, individuals can protect their brain health and reduce their risk of developing Alzheimer's disease.

Q17: How does education and lifelong learning impact Alzheimer's risk?

Education and lifelong learning have a significant impact on Alzheimer's risk by enhancing cognitive reserve. Cognitive reserve refers to the brain's ability to compensate for age-related changes and neurodegenerative processes. Higher levels of education and continuous mental stimulation throughout life can promote the development of neural connections and brain plasticity. Engaging in lifelong learning activities, such as reading, attending classes, learning new skills, and participating in intellectually stimulating hobbies, can help maintain cognitive function and delay the onset of Alzheimer's symptoms. Research has shown that individuals with higher levels of education and those who engage in regular cognitive activities have a lower risk of developing Alzheimer's disease. By staying mentally active and continuously challenging the brain, individuals can build cognitive reserves and reduce their risk of cognitive decline and Alzheimer's.

Q18: How do genetic factors influence Alzheimer's risk, and can lifestyle changes still make a difference?

Genetic factors, such as the presence of the APOE ε4 allele, can increase an individual's risk of developing Alzheimer's disease. However, having a genetic predisposition does not guarantee the development of Alzheimer's, and lifestyle changes can still make a significant difference. While genetic factors are beyond one's control, adopting a healthy lifestyle can help mitigate the risk. Regular physical exercise, a balanced diet, cognitive engagement, social interaction, good sleep hygiene, and stress management can all contribute to overall brain health and potentially delay the onset of Alzheimer's symptoms. Additionally, managing other risk factors such as hypertension, diabetes, and cardiovascular health is crucial. Even individuals with a genetic predisposition to Alzheimer's can benefit from a proactive approach to brain health and reduce their overall risk through lifestyle modifications.

Q19: What are the benefits of mindfulness and meditation in Alzheimer's prevention?

Mindfulness and meditation offer several benefits in Alzheimer's prevention by reducing stress, enhancing emotional well-being, and improving cognitive function. Chronic stress and elevated levels of cortisol can damage brain cells and impair memory and learning. Mindfulness and meditation practices help lower cortisol levels, reduce stress, and promote relaxation. These practices also improve attention, concentration, and emotional regulation, all of which are important for cognitive health. Regular mindfulness and meditation can enhance brain plasticity and increase gray matter in brain regions associated with memory and executive function. Studies have shown that individuals who practice mindfulness and meditation regularly have better cognitive function and a lower risk of cognitive decline. Incorporating mindfulness and meditation into daily routines can support overall brain health and reduce the risk of Alzheimer's disease.

Q20: How does managing other chronic health conditions influence Alzheimer's risk?

Managing other chronic health conditions, such as hypertension, diabetes, cardiovascular disease, and obesity, is crucial for reducing Alzheimer's risk. These conditions can negatively impact brain health by reducing blood flow, increasing inflammation, and contributing to metabolic dysfunction. Effective management of these conditions through medication, lifestyle changes, and regular medical check ups can help reduce these risks. For example, controlling blood pressure and blood sugar levels, maintaining a healthy weight, and adopting a heart-healthy diet can improve overall metabolic health and protect the brain from vascular damage. By managing chronic health conditions effectively, individuals can reduce their risk of cognitive decline and support overall brain health, thereby lowering the likelihood of developing Alzheimer's disease.

Q21: How do genetic factors influence the risk of Alzheimer's disease, and what can individuals with a genetic predisposition do to mitigate this risk?

Genetic factors play a significant role in the risk of developing Alzheimer's disease. One of the most well-known genetic risk factors is the presence of the APOE E4 allele. Individuals who inherit one copy of this allele have an increased risk of developing Alzheimer's, while those with two copies have an even higher risk. However, having the APOE E4 allele does not guarantee that a person will develop Alzheimer's; it simply raises the likelihood compared to those without the allele. There are other genetic mutations, such as those in the APP, PSEN1, and PSEN2 genes, which are associated with early-onset familial Alzheimer's disease, a much rarer form that typically begins before age 65.

Despite the influence of genetics, individuals with a genetic predisposition to Alzheimer's can take proactive steps to mitigate their risk. Lifestyle modifications can significantly impact brain health and potentially delay the onset of Alzheimer's symptoms. Regular physical exercise is one of the most effective strategies, as it enhances cardiovascular health, increases blood flow to the brain, and supports the growth of new neurons. A healthy diet, such as the Mediterranean diet, which is rich in fruits, vegetables, whole grains, and healthy fats, can reduce inflammation and provide essential nutrients for brain function.

Cognitive engagement is also crucial. Activities that challenge the brain, such as puzzles, reading, and learning new skills, can build cognitive reserve, making the brain more resilient to the effects of neurodegeneration. Social interaction and maintaining strong social networks can help reduce stress and depression, both of which are risk factors for Alzheimer's. Additionally, managing other health conditions such as hypertension, diabetes, and high cholesterol is essential. These conditions can exacerbate the risk of Alzheimer's, so controlling them through medication, diet, and lifestyle changes is vital.

In summary, while genetic factors can increase the risk of Alzheimer's, individuals can adopt various lifestyle changes to protect their brain health and potentially reduce their risk. These proactive measures are valuable for everyone, but especially for those with a genetic predisposition to Alzheimer's.

In conclusion, while Alzheimer's disease remains a formidable challenge with no current cure, the advancement in understanding its prevention and risk reduction offers a beacon of hope. Emphasizing a proactive approach through lifestyle modifications can significantly alter the trajectory of the disease. Regular physical activity, which promotes cardiovascular health and neuroplasticity, and a balanced diet, rich in antioxidants and healthy fats, are foundational elements in this preventive strategy. Cognitive engagement through lifelong learning and mental exercises helps build cognitive reserve, a crucial defense against the progression of Alzheimer's. Additionally, fostering social connections not only enhances mental well-being but also provides the necessary emotional support to combat stress and depression, both known risk factors for cognitive decline. Adequate sleep, often overlooked, plays a vital role in clearing brain toxins and supporting memory consolidation. Stress management, through mindfulness, meditation, or other relaxation techniques, helps mitigate the harmful effects of chronic stress on brain health. Furthermore, managing chronic conditions like hypertension, diabetes, and obesity through medical intervention and lifestyle changes can significantly lower the risk of Alzheimer's. It is essential for public health policies to integrate these preventive measures into community health programs, ensuring accessibility and education for all demographic groups. By adopting a comprehensive and informed approach to Alzheimer's prevention, we can not only improve individual health outcomes but also alleviate the broader societal impact of this debilitating disease. As research continues to evolve, the integration of scientific discoveries into practical, everyday practices remains pivotal. Ultimately, through concerted efforts and a commitment to healthier lifestyles, we can aspire to a future where the incidence of Alzheimer's is significantly reduced, offering hope and improved quality of life for aging populations globally.

Bibliography

1. ***Anstey, K. J., von Sanden, C., Salim, A., & O'Kearney, R. (2007).*** *Smoking as a risk factor for dementia and cognitive decline: A meta-analysis of prospective studies. American Journal of Epidemiology, 166(4), 367–378. https://doi.org/10.1093/aje/kwm116*
2. ***Barnes, D. E., & Yaffe, K. (2011).*** *The projected effect of risk factor reduction on Alzheimer's disease prevalence. The Lancet Neurology, 10(9), 819–828. https://doi.org/10.1016/S1474-4422(11)70072-2*
3. ***Diniz, B. S., Butters, M. A., Albert, S. M., Dew, M. A., & Reynolds, C. F. (2013).*** *Late-life depression and risk of vascular dementia and Alzheimer's disease: Systematic review and meta-analysis of community-based cohort studies. The British Journal of Psychiatry, 202(5), 329–335. https://doi.org/10.1192/bjp.bp.112.118307*
4. ***Innes, K. E., & Selfe, T. K. (2014).*** *Meditation and music improve memory and cognitive function in adults with subjective cognitive decline: A pilot randomized controlled trial. Journal of Alzheimer's Disease, 52(4), 1277–1288. https://doi.org/10.3233/JAD-150254*
5. ***Livingston, G., Sommerlad, A., Orgeta, V., Costafreda, S. G., Huntley, J., Ames, D., ... & Mukadam, N. (2017).*** *Dementia prevention, intervention,*

and care. The Lancet, 390(10113), 2673–2734. https://doi.org/10.1016/S0140-6736(17)31363-6

6. ***Luchsinger, J. A., Patel, B., Tang, M. X., Schupf, N., & Mayeux, R. (2007).*** *Measures of adiposity and dementia risk in elderly persons. Archives of Neurology, 64(3), 392–398. https://doi.org/10.1001/archneur.64.3.392*
7. ***Solomon, A., Kivipelto, M., Wolozin, B., Zhou, J., & Whitmer, R. A. (2009).*** *Midlife serum cholesterol and increased risk of Alzheimer's and vascular dementia three decades later. Dementia and Geriatric Cognitive Disorders, 28(1), 75–80. https://doi.org/10.1159/000232589*
8. ***Valenzuela, M. J., & Sachdev, P. (2006).*** *Brain reserve and dementia: A systematic review. Psychological Medicine, 36(4), 441–454. https://doi.org/10.1017/S0033291705006264*

CHAPTER 14

Policy and Advocacy for Alzheimer's Disease

In the global fight against Alzheimer's disease, policy and advocacy serve as crucial pillars shaping research, care, and support initiatives. In this chapter we will be focusing on "Policy and Advocacy for Alzheimer's Disease," from legislative mandates to grassroots movements, efforts to address Alzheimer's at the policy level are integral to advancing understanding, improving care outcomes, and ultimately finding a cure for this devastating illness.

Q1: What is the role of policy in addressing the challenges posed by Alzheimer's disease?

Policy plays a pivotal role in addressing the multifaceted challenges posed by Alzheimer's disease by providing a framework for comprehensive approaches to prevention, treatment, care, and support. Alzheimer's disease is not only a medical issue but also a social, economic, and public health challenge that requires coordinated efforts across various sectors. Policies aimed at Alzheimer's disease typically encompass several key areas: research funding, healthcare delivery, caregiver support, public education, and legal protections for individuals with dementia.

At the heart of effective Alzheimer's policy is the allocation of resources towards research initiatives that aim to better understand the disease's underlying mechanisms, identify biomarkers for early detection, and develop disease-modifying treatments. Policymakers collaborate with scientists, healthcare providers, and advocacy groups for advancing Alzheimer's care and ultimately finding a cure. National Alzheimer's plans, such as the National Alzheimer's Project Act (NAPA) in the United States, mandate strategic goals and objectives to guide these efforts, ensuring that research funding and policy initiatives are aligned with national priorities.

Beyond research, Alzheimer's policies address the healthcare needs of individuals living with the disease and their caregivers. They promote access to timely diagnosis, specialized care services, and support programs that enhance quality of life. This includes dementia-specific training for healthcare professionals to improve diagnostic accuracy and provide compassionate care tailored to the unique needs of Alzheimer's patients.

Caregivers, who play a crucial role in supporting individuals with Alzheimer's, benefit from policies that offer respite care, financial assistance, and workplace accommodations. Policies also aim to protect the legal rights of individuals with dementia, ensuring that they receive adequate representation and support in decision-making processes related to their healthcare and financial affairs. Public education and awareness campaigns are another cornerstone of Alzheimer's policy, aimed at reducing stigma, promoting early detection, and encouraging healthy lifestyle choices that may reduce the risk of cognitive decline. These campaigns foster community understanding and support for individuals affected by Alzheimer's, driving societal change and advocacy efforts at local, national, and international levels.

In summary, Alzheimer's policies are essential for coordinating efforts across sectors, mobilizing resources, and advocating for the rights and well-being of individuals with Alzheimer's disease and their caregivers. By addressing the complex challenges of Alzheimer's through strategic policy interventions, governments and organizations can make significant strides towards improving care, advancing research, and ultimately finding a cure for this devastating disease.

Q2: How does advocacy influence Alzheimer's research funding and priorities?

Advocacy plays a pivotal role in influencing Alzheimer's research funding and priorities by mobilizing public support, raising awareness among policymakers, and advocating for increased investment in research initiatives. Alzheimer's advocacy organizations and grassroots movements amplify the voices of individuals affected by the disease, compelling policymakers to prioritize Alzheimer's research as a critical public health issue.

One of the primary ways advocacy influences research funding is by advocating for increased government funding through legislative efforts. Advocacy organizations work closely with lawmakers to secure appropriations for agencies like the National Institutes of Health (NIH) and the National Institute on Aging (NIA), which fund a significant portion of Alzheimer's research in the United States. These funds support a wide range of research initiatives, from basic science to clinical trials aimed at developing new treatments and interventions.

Advocacy efforts also focus on shaping research priorities to address gaps in knowledge and focus on areas with the greatest potential for impact. For example, advocates may push for increased funding for research on early detection methods, novel biomarkers, or non-pharmacological interventions that improve quality of life for Alzheimer's patients.

Beyond government funding, advocacy can influence private-sector investment in Alzheimer's research by highlighting the societal and economic impact of the disease. By demonstrating the potential return on investment in Alzheimer's research, advocates encourage pharmaceutical companies and biotech firms to allocate resources towards developing innovative treatments and therapies. Moreover, advocacy organizations collaborate with researchers, healthcare providers, and industry stakeholders to ensure that research efforts are collaborative and interdisciplinary. They facilitate partnerships that accelerate scientific discoveries, promote data sharing, and translate research findings into clinical practice.

In summary, advocacy influences Alzheimer's research funding and priorities by mobilizing public support, engaging policymakers, and fostering collaboration among stakeholders. By advocating for increased investment in research and shaping research priorities, advocates play a critical role in advancing scientific knowledge, improving care outcomes, and ultimately finding effective treatments and a cure for Alzheimer's disease.

Q3: What are some key legislative initiatives aimed at addressing Alzheimer's disease?

Key legislative initiatives aimed at addressing Alzheimer's disease include landmark acts and policies designed to enhance research, care, and support for individuals affected by the disease. In the United States, the National Alzheimer's Project Act (NAPA) stands out as a comprehensive legislative effort to tackle Alzheimer's at a national level. Enacted in 2011, NAPA mandated the development and implementation of a national plan to address Alzheimer's and related dementias. The plan outlines strategic goals and objectives across several key domains, including research, clinical care, caregiver support, public awareness, and data infrastructure. NAPA has catalyzed efforts to coordinate federal resources, streamline research funding, and promote collaboration among stakeholders from government, academia, industry, and advocacy organizations.

Internationally, countries have implemented similar initiatives tailored to their healthcare systems and societal needs. For instance, countries in the European Union have adopted strategies under the Joint Programming Initiative on Neurodegenerative Diseases (JPND), which coordinates research efforts across member states to advance understanding of Alzheimer's disease and develop innovative treatments. These initiatives reflect a global commitment to addressing the Alzheimer's crisis through coordinated policy measures and international collaboration.

Legislative efforts also extend to funding mechanisms that support Alzheimer's research and care. In addition to direct appropriations for research agencies like the National Institutes of Health (NIH) and the Centers for Disease Control and Prevention (CDC), legislative bodies may establish grant programs, tax incentives for research and development, and public–private partnerships that incentivize innovation in Alzheimer's science and healthcare delivery.

Moreover, legislative initiatives often include provisions for enhancing caregiver support and workforce training. Policies may provide funding for caregiver respite services, training programs for healthcare professionals on dementia care, and initiatives to promote dementia-friendly communities. These efforts aim to improve the quality of life for individuals living with Alzheimer's and their caregivers, reduce caregiver burden, and ensure that healthcare providers are equipped to deliver compassionate and effective care.

Overall, legislative initiatives aimed at addressing Alzheimer's disease represent a multifaceted approach to tackling one of the greatest public health challenges of our time. By enacting comprehensive policies that prioritize research, enhance care and support, and promote public awareness, policymakers can make significant strides towards improving outcomes for individuals affected by Alzheimer's and advancing the search for effective treatments and a cure.

Q4: How do national Alzheimer's plans contribute to tackling the disease?

National Alzheimer's plans are strategic frameworks that guide coordinated efforts to tackle Alzheimer's disease on a national scale. These plans outline goals, objectives, and actionable strategies across key domains such as research, clinical care, caregiver support, public awareness, and data infrastructure. By setting clear priorities and aligning resources, national Alzheimer's plans facilitate collaboration among stakeholders from government agencies, research institutions, healthcare providers, advocacy organizations, and the private sector.

A primary focus of national Alzheimer's plans is advancing research to better understand the underlying causes of Alzheimer's disease, identify biomarkers for early detection, and develop new treatments and interventions. Plans allocate funding to support basic science, clinical trials, and translational research efforts aimed at accelerating scientific discoveries and bringing promising therapies to market.

National Alzheimer's plans also emphasize improving clinical care and enhancing support services for individuals living with Alzheimer's and their caregivers. This includes promoting early detection through screening programs, training healthcare providers on dementia care best practices, and expanding access to specialized services such as memory clinics and dementia care centers.

Caregiver support is another critical component addressed by national Alzheimer's plans. Policies may include funding for caregiver respite services, educational programs on caregiving skills, and initiatives to promote caregiver health and well-being. By reducing caregiver burden and enhancing support networks, these efforts aim to improve quality of life for caregivers and enable them to provide better care for their loved ones with Alzheimer's. Public awareness and education campaigns are integral to national Alzheimer's plans, aiming to reduce stigma, promote early diagnosis, and educate the public about risk factors and preventive measures. These campaigns engage media, community organizations, and healthcare professionals to raise awareness, foster understanding, and encourage proactive steps towards brain health and Alzheimer's prevention. Data infrastructure is another key area addressed by national Alzheimer's plans, focusing on improving data collection, sharing, and analysis to inform policy decisions, track disease trends, and evaluate the impact of interventions. By establishing robust data systems, policymakers can monitor progress, identify gaps, and allocate resources effectively to address evolving challenges posed by Alzheimer's disease.

In summary, national Alzheimer's plans provide a comprehensive roadmap for addressing the complex challenges of Alzheimer's disease through coordinated research, clinical care, caregiver support, public awareness, and data infrastructure initiatives. By leveraging collaborative efforts and aligning resources towards common goals, these plans play a crucial role in advancing scientific knowledge, improving care outcomes, and ultimately finding effective treatments and a cure for Alzheimer's.

Q5: What is the economic impact of Alzheimer's disease, and how does policy address this issue?

Alzheimer's disease exerts a significant economic impact on individuals, families, healthcare systems, and society at large due to direct medical costs, long-term care expenses, lost productivity, and caregiver burden. As the global population ages, the economic burden of Alzheimer's is projected to escalate, highlighting the urgent need for effective policy measures to mitigate these costs and improve care outcomes.

The economic impact of Alzheimer's disease spans multiple sectors, including healthcare, social services, and the labor market. Direct medical costs associated with Alzheimer's include expenses for hospitalizations, medications, physician visits, and long-term care services such as nursing home care and home healthcare. Indirect costs arise from lost productivity due to disability, premature retirement, and absenteeism among caregivers who often face challenges balancing work and caregiving responsibilities.

Policy initiatives aimed at addressing the economic impact of Alzheimer's disease focus on several key areas. One critical aspect is enhancing healthcare delivery and reimbursement policies to ensure access to timely diagnosis, specialized care, and evidence-based treatments. Policymakers may advocate for Medicare and Medicaid reforms that support comprehensive dementia care services, including coverage for cognitive assessments, caregiver support programs, and care coordination services.

Long-term care financing is another area of focus for Alzheimer's policy, as individuals with the disease often require extensive support over many years. Policies may include incentives for long-term care insurance, Medicaid eligibility expansions for home- and community-based services, and tax incentives for caregivers to offset caregiving expenses. These initiatives aim to alleviate financial strain on families while ensuring access to quality care options that promote independence and dignity for individuals with Alzheimer's.

Research funding policies also play a crucial role in addressing the economic impact of Alzheimer's disease by investing in innovative research initiatives aimed at developing disease-modifying treatments and preventive interventions. By advancing scientific knowledge and accelerating the development of effective therapies, policymakers seek to reduce long-term healthcare costs associated with Alzheimer's and improve health outcomes for affected individuals.

Furthermore, policies that support caregiver workforce participation and provide workplace accommodations for caregivers can mitigate lost productivity and economic hardship. Initiatives may include caregiver leave policies, flexible work arrangements, and employer-sponsored caregiver support programs that promote work–life balance and retention in the workforce.

In summary, addressing the economic impact of Alzheimer's disease requires comprehensive policy solutions that span healthcare delivery, long-term care financing, research funding, and support for caregivers. By implementing effective policies, policymakers can enhance care quality, reduce financial burden on families, and promote economic productivity while advancing towards the ultimate goal of finding a cure for Alzheimer's disease.

Q6: How can policy support caregivers of individuals with Alzheimer's disease?

Policy plays a critical role in supporting caregivers of individuals with Alzheimer's disease by providing resources, services, and protections that alleviate caregiver burden and improve quality of life. Caregivers play a pivotal role in the care and support of individuals with Alzheimer's, often facing significant physical, emotional, and financial challenges. Policy initiatives aim to recognize and address these challenges through various strategies.

One key aspect of caregiver support policy is the provision of respite care services, which offer temporary relief to caregivers by providing substitute care for their loved ones. Respite care programs may include in-home care services, adult day programs, and short-term residential stays, allowing caregivers to take breaks, attend to their own health needs, or engage in personal activities without compromising the quality of care provided to the individual with Alzheimer's.

Financial assistance programs are another critical component of caregiver support policy, providing financial relief to offset the costs associated with caregiving. These programs may include tax credits for caregiving expenses, caregiver stipends through Medicaid waiver programs, and subsidies for long-term care services. By alleviating financial strain, caregivers can better focus on providing compassionate care and enhancing the quality of life for individuals with Alzheimer's.

Policies that promote caregiver training and education are essential for equipping caregivers with the knowledge and skills needed to provide effective care. Training programs may cover topics such as dementia care techniques, communication strategies, behavior management, and self-care practices. By enhancing caregiver competence and confidence, these programs improve caregiving outcomes and reduce stress for both caregivers and individuals with Alzheimer's.

Workplace policies that support caregivers are crucial for balancing work responsibilities with caregiving duties. Flexible work arrangements, caregiver leave policies, and employee assistance programs can help caregivers manage their professional commitments while attending to the needs of their loved ones with Alzheimer's. These policies promote work–life balance, reduce absenteeism, and support retention of experienced caregivers in the workforce.

Legal protections for caregivers are also addressed by policy initiatives, ensuring that caregivers have the necessary legal authority to make healthcare decisions, manage financial affairs, and access medical information on behalf of individuals with Alzheimer's. Policies may include durable power of attorney, guardianship provisions, and healthcare proxy designations that safeguard the rights and well-being of both caregivers and care recipients.

Q7: What role do international collaborations play in addressing Alzheimer's disease?

International collaborations play a crucial role in addressing Alzheimer's disease by fostering global cooperation, sharing scientific knowledge, harmonizing research efforts, and promoting best practices in care and treatment. Alzheimer's disease is a global health challenge that transcends national boundaries, requiring coordinated efforts and collective action to advance understanding and develop effective solutions.

One key aspect of international collaborations is the sharing of scientific research and data among countries and research institutions. Collaborative networks facilitate the exchange of findings on Alzheimer's genetics, biomarkers, epidemiology, and clinical trials, enabling researchers to build upon each other's work and accelerate scientific progress. By pooling resources and expertise, international collaborations enhance the efficiency and impact of research efforts aimed at unraveling the complexities of Alzheimer's disease.

Joint research initiatives and consortia play a pivotal role in international collaborations by coordinating large-scale studies, multi center clinical trials, and longitudinal cohorts that span diverse populations and geographic regions. These initiatives enable researchers to investigate genetic variations, environmental factors, and socio cultural influences on Alzheimer's risk and progression, generating insights that are applicable across global contexts.

International collaborations also promote the development of standardized diagnostic criteria, treatment guidelines, and care protocols for Alzheimer's disease. Consensus-based approaches ensure consistency in clinical practice and enhance healthcare delivery worldwide, improving diagnostic accuracy, treatment outcomes, and quality of life for individuals living with Alzheimer's.

Policy harmonization is another critical area addressed by international collaborations, as countries work together to align regulatory frameworks, ethical standards, and data privacy protections for Alzheimer's research and clinical trials. By establishing common guidelines and protocols, international collaborations facilitate the efficient and ethical conduct of research while safeguarding participant rights and promoting transparency.

Furthermore, international collaborations enhance advocacy efforts by amplifying the collective voice of stakeholders, including advocacy organizations, healthcare professionals, policymakers, and individuals affected by Alzheimer's. Collaborative advocacy campaigns raise global awareness, mobilize public support, and influence policy decisions that prioritize Alzheimer's research funding, caregiver support, and healthcare reform on a global scale.

In summary, international collaborations are essential for addressing Alzheimer's disease by promoting global cooperation, sharing scientific knowledge, harmonizing research efforts, and advancing policy initiatives. By working together across borders and disciplines, countries can accelerate progress towards understanding the causes of Alzheimer's, developing effective treatments, and ultimately finding a cure for this devastating disease.

Q8: How do policies ensure equitable access to Alzheimer's care and services?

Policies play a crucial role in ensuring equitable access to Alzheimer's care and services by addressing disparities in healthcare access, promoting inclusion, and supporting initiatives that reach underserved populations. Alzheimer's disease affects individuals from diverse backgrounds, yet access to timely diagnosis, specialized care, and support services can vary significantly based on factors such as geography, socioeconomic status, and cultural considerations. One approach taken by policymakers is the development of dementia-friendly communities and healthcare systems that accommodate the needs of individuals living with Alzheimer's and their caregivers. Dementia-friendly initiatives may include training healthcare professionals on culturally competent care practices, adapting clinical settings to be more accessible, and providing language interpretation services to facilitate communication and understanding.

Financial barriers to Alzheimer's care are addressed through policies that expand coverage options and reimbursement mechanisms for diagnostic testing, medical treatments, and long-term care services. Policymakers may advocate for Medicaid eligibility expansions, Medicare coverage improvements, and tax incentives for families to offset caregiving expenses, ensuring that individuals with Alzheimer's receive the comprehensive care they need without financial hardship.

Community-based programs and support services are another focus of Alzheimer's care policies, aiming to enhance access to respite care, caregiver support groups, transportation assistance, and home-based healthcare services. These initiatives promote independence, social engagement, and quality of life for individuals with Alzheimer's and their caregivers, reducing reliance on institutional care and supporting aging in place.

Educational outreach and public awareness campaigns are integral to policies promoting equitable access to Alzheimer's care. By raising awareness about Alzheimer's risk factors, symptoms, and available resources, policymakers empower communities to recognize early signs of cognitive decline, seek timely medical attention, and access supportive services that promote better outcomes and quality of life.

Furthermore, policies may include incentives for healthcare providers to offer comprehensive dementia care services, such as memory clinics, interdisciplinary care teams, and personalized care plans tailored to individual needs. By promoting continuity of care and care coordination, these initiatives ensure that individuals with Alzheimer's receive holistic, patient-centered care that addresses their physical, emotional, and cognitive well-being.

In summary, policies that ensure equitable access to Alzheimer's care and services address healthcare disparities, promote inclusion, and support initiatives that reach underserved populations. By expanding coverage options, enhancing community-based support services, and raising public awareness, policymakers can improve care outcomes, reduce disparities, and enhance quality of life for individuals affected by Alzheimer's disease.

Q9: What are some challenges in advocating for Alzheimer's disease at the policy level?

Advocating for Alzheimer's disease at the policy level presents several challenges, including competing healthcare priorities, budget constraints, stigma, and the complex nature of Alzheimer's as a multifaceted disease that requires comprehensive strategies and long-term commitments.

One significant challenge is competing healthcare priorities within legislative agendas and budget allocations. Policymakers must prioritize funding and resources across a range of healthcare issues, from infectious diseases and chronic conditions to mental health and aging-related challenges. Advocates for Alzheimer's must demonstrate the urgency and societal impact of the disease to secure sufficient funding and support for research, care, and support services.

Budget constraints pose another challenge, as governments and healthcare systems face financial pressures that limit investment in Alzheimer's research, prevention programs, and caregiver support initiatives. Advocates must make a compelling case for the economic and social benefits of investing in Alzheimer's, emphasizing the potential cost savings associated with early intervention, improved care outcomes, and reduced long-term care expenditures.

Stigma associated with Alzheimer's disease and dementia can hinder advocacy efforts by perpetuating misconceptions, fear, and reluctance to discuss cognitive decline openly. Advocates work to raise awareness, challenge stereotypes, and promote compassionate understanding of Alzheimer's as a medical condition that deserves attention, research investment, and support for affected individuals and their families.

The complex nature of Alzheimer's disease presents challenges in developing and implementing effective policy solutions that address diverse needs across the continuum of care. Alzheimer's is not a singular condition but rather a spectrum of cognitive impairments that vary in severity, progression, and underlying causes. Policies must be flexible, inclusive, and responsive to evolving scientific discoveries and healthcare innovations.

Moreover, advocacy efforts may encounter resistance from vested interests, ideological differences, and bureaucratic hurdles within governmental agencies and healthcare systems. Building consensus, fostering bipartisan support, and engaging stakeholders from diverse sectors are essential strategies for advancing Alzheimer's advocacy goals and overcoming resistance to policy change.

Despite these challenges, advocates leverage public awareness campaigns, grassroots mobilization, and strategic alliances with policymakers, researchers, healthcare providers, and community organizations to drive momentum for Alzheimer's policy reform. By raising voices, sharing personal stories, and advocating for meaningful change, stakeholders can influence policy decisions, secure funding, and improve outcomes for individuals affected by Alzheimer's disease and their caregivers.

In summary, advocating for Alzheimer's disease at the policy level requires overcoming challenges such as competing priorities, budget constraints, stigma, and the complexity of the disease. By addressing these challenges through strategic advocacy efforts, collaboration, and persistence, advocates can advance policy reforms that enhance research, care, and support for individuals living with Alzheimer's and their families.

Q10: How can policy promote research into new treatments and therapies for Alzheimer's disease?

Policy plays a critical role in promoting research into new treatments and therapies for Alzheimer's disease by allocating funding, incentivizing innovation, and streamlining regulatory processes that accelerate the development and approval of effective treatments.

One of the primary ways policy promotes Alzheimer's research is through government funding initiatives that support basic science, translational research, and clinical trials. Agencies such as the National Institutes of Health (NIH), the National Institute on Aging (NIA), and the Alzheimer's Association allocate significant resources to research programs that investigate the underlying mechanisms of Alzheimer's, identify potential therapeutic targets, and test novel interventions aimed at slowing or halting disease progression. Policymakers collaborate with research institutions, pharmaceutical companies, advocacy organizations, and academic centers to prioritize funding for high-impact research projects that hold promise for advancing Alzheimer's treatment options. By fostering collaboration and innovation, policymakers create opportunities for interdisciplinary research teams to explore diverse approaches, from drug discovery and gene therapy to lifestyle interventions and digital health technologies.

In addition to funding, policy initiatives include incentives for private-sector investment in Alzheimer's research and development. Tax credits, research grants, and public–private partnerships incentivize pharmaceutical companies, biotech firms, and venture capitalists to invest in innovative therapies and technologies that address unmet medical needs in Alzheimer's care.

Regulatory policies also play a crucial role in promoting research by streamlining approval processes for Alzheimer's drugs and therapies. Policymakers work with regulatory agencies such as the Food and Drug Administration (FDA) to expedite review timelines, establish clear guidelines for clinical trial design, and incorporate patient-centered outcomes into regulatory decision-making. These efforts reduce barriers to innovation, encourage early-stage investments in Alzheimer's research, and facilitate the translation of scientific discoveries into clinical practice. Furthermore, policies that promote data sharing, research transparency, and collaboration across international borders enhance the efficiency and impact of Alzheimer's research efforts. By facilitating the exchange of research findings, biomarker data, and clinical trial outcomes, policymakers enable researchers to build upon each other's work, validate findings, and accelerate the development of effective treatments. Public–private partnerships are another key strategy for advancing Alzheimer's research, as policymakers collaborate with industry stakeholders to leverage expertise, resources, and infrastructure for large-scale research initiatives. These partnerships promote precompetitive research collaborations, data sharing agreements, and joint investment in innovative research platforms that accelerate progress towards breakthrough treatments for Alzheimer's disease.

In summary, policy promotes research into new treatments and therapies for Alzheimer's disease by allocating funding, incentivizing innovation, streamlining regulatory processes, and fostering collaboration among stakeholders. By prioritizing research investment, reducing barriers to drug development, and promoting public–private partnerships, policymakers can accelerate scientific progress, improve care outcomes, and ultimately find effective treatments and a cure for Alzheimer's disease.

As we navigate the complexities of Alzheimer's disease, ongoing advocacy and informed policymaking remain essential. By fostering collaboration, raising awareness, and prioritizing research funding, we can collectively work towards a future where effective treatments and compassionate care are accessible to all affected by Alzheimer's. Together, through sustained advocacy efforts, we can make a meaningful difference in the lives of individuals with Alzheimer's and their families, ensuring dignity, support, and hope for the future.

Bibliography

1. ***Alzheimer's Association. (2022).*** *Advocacy and public policy. https://www.alz.org/advocacy*
2. ***Alzheimer's Association. (2022).*** *Costs of Alzheimer's and dementia. https://www.alz.org/alzheimers-dementia/facts-figures*
3. ***Alzheimer's Association. (2022).*** *Public policy positions. https://www.alz.org/publicpolicy*
4. ***Alzheimer's Impact Movement. (2022).*** *Advocacy priorities. https://alzimpact.org/advocacy-priorities/*
5. ***Alzheimer's Society. (2022).*** *Dementia-friendly communities. https://www.alzheimers.org.uk/get-support/dementia-friendly-communities*
6. ***Centers for Medicare & Medicaid Services. (2022).*** *Alzheimer's disease care planning. https://www.cms.gov/*
7. ***European Union Joint Programming Initiative on Neurodegenerative Diseases (JPND). (2022).*** *About JPND. https://www.jpnd.eu/*

8. ***Family Caregiver Alliance. (2022).*** *Alzheimer's caregiving. https://www.caregiver.org/caregiving-alzheimers-disease-or-dementia*
9. ***National Institute on Aging. (2022).*** *Alzheimer's disease research summit 2022. https://www.nia.nih.gov/news/events/2022/06/alzheimers-disease-research-summit-2022*
10. ***National Institute on Aging. (2022).*** *National Alzheimer's project act (NAPA). https://www.congress.gov/bill/111th-congress/house-bill/4689*
11. ***U.S. Congress. (2011).*** *National Alzheimer's project act (NAPA). https://www.congress.gov/bill/111th-congress/house-bill/4689*
12. ***U.S. Department of Health and Human Services. (2022).*** *National plan to address Alzheimer's disease. https://aspe.hhs.gov/national-plans-address-alzheimers-disease*
13. ***U.S. Department of Health and Human Services. (2022).*** *Support for caregivers. https://www.hhs.gov/aging/index.html*
14. ***U.S. Food and Drug Administration. (2022).*** *Alzheimer's disease drug development. https://www.fda.gov/*
15. ***World Health Organization. (2022).*** *Dementia: Key facts. https://www.who.int/news-room/fact-sheets/detail/dementia*

Conclusion: Promising Paths Forward

Alzheimer's disease presents complex challenges that require multifaceted approaches to research, treatment, and caregiving. Despite the current lack of disease-modifying treatments, advances in understanding Alzheimer's pathology, biomarker development, and precision medicine offer hope for future interventions that could slow or prevent the progression of the disease. Non-pharmacological therapies, including cognitive stimulation, physical exercise, and social engagement, play important roles in managing symptoms and improving quality of life for individuals with Alzheimer's. Caregivers and healthcare providers are essential partners in the journey of Alzheimer's care, and supporting their well-being is crucial for providing high-quality, person-centered care. As the global burden of Alzheimer's continues to grow, concerted efforts in research, advocacy, and public policy are needed to address the societal impacts of the disease and ensure that individuals with Alzheimer's and their families receive the support and resources they need. By working together and embracing innovation, we can advance our understanding of Alzheimer's disease and ultimately transform the landscape of dementia care for the better.

Index

B

D

E

F

G

H

I

J

K

L

M

www.ingramcontent.com/pod-product-compliance
Lightning Source LLC
LaVergne TN
LVHW080040210926
848481LV00011B/3246
9781501527463